Building Research Establishment Report

Thermal insulation: avoiding risks

A good practice guide supporting building regulations requirements

Building Research Establishment
Garston
Watford
WD2 7JR

London: HMSO

Prices for all available
BRE publications can be
obtained from:

BRE Bookshop
Building Research Establishment
Garston, Watford, WD2 7JR
Telephone: 0923 664444
Fax: 0923 664400

The guidance in this report draws upon the research and experience of the Building Research Establishment. It has been subjected to public comment and amended wherever possible to reflect a consensus of views. It was prepared for publication by NBA Tectonics, under the guidance of BRE staff.

This book is published jointly by the
Building Research Establishment as BR 262 ISBN 0 85125 632 5
and
HMSO ISBN 0 11 701792 2

© Crown copyright 1994
Second edition 1994
First edition Crown copyright 1989

Applications to copy all or any part of
this publication should be made to
the Publications Manager
at the Building Research Establishment

Introduction

SCOPE AND FORMAT

This book is an extensive revision of a guide first produced in 1989. The format has been altered to improve the link between risks, causes and solutions. It also contains many new illustrations. Results of research, changes in materials and new construction techniques have all been incorporated. The scope has been extended to include more information on non-domestic buildings, the conversion or alteration of existing buildings and ways of reducing air infiltration, and thereby heat losses.

The guide has been prepared to support building regulations for the conservation of fuel and power. The information in the guide represents the recommendations of BRE on good design and construction practice associated with thermal standards; it is not a document approved to satisfy all requirements of building regulations. Users of the publication are responsible for the correct application of the advice.

The guide discusses the more important technical risks associated with meeting the requirements of building regulations for thermal insulation. Technical risks are highlighted by a pink tint and these are followed by actions that could be taken to avoid the risk. In assessing risks for a particular building, consideration should be given to the environmental conditions likely to occur both inside and outside the building, and its expected life.

The guide is divided into five sections relating to the major elements of the building. These sections are broken down into sub-sections to reflect alternative construction methods and the impact of the position of insulation within the construction. Each section concludes with a list of quality control checks for use on site.

Illustrations outline construction principles and good practice details. These examples are not exhaustive and designers and builders may have established other details that are equally suitable. For example, in many circumstances partial cavity insulation may be used instead of full cavity insulation, and vice versa.

Throughout the guide illustrations and recommended actions are cross-referenced.

A new section which considers the effects of improved insulation on the building as a whole has been introduced. It outlines the interaction between thermal insulation, ventilation and heating in avoiding potential risks and gives outline guidance on factors to be considered.

 Clauses marked thus apply to buildings subject to renovation, conversion or alteration.

DEFINITIONS

Breather membrane
A thin membrane with a vapour resistance in the range 0.1 to 2.0 MNs/g — used for restricting liquid water penetration whilst allowing water vapour transfer, eg as a tiling underlay or covering to sheathing.

Vapour control layer
A term which replaces both vapour check and vapour barrier for a continuous or sealed/jointed membrane made of materials with a vapour resistance greater than 200 MNs/g — used for restricting water vapour flow into the construction.

Low permeability insulation
A rigid insulation board with a vapour resistivity greater than 200 MNs/gm — used where it is advantageous to restrict the amount of water vapour transfer through the insulation layer.

Insulating blockwork
Masonry with blocks having a thermal conductivity no greater than 0.3 W/mK — used to reduce thermal bridging.

Contents

Introduction

1 Whole building
1

2 Roofs
Pitched roofs with ventilated roof spaces — 3
Pitched roofs with sarking insulation — 8
Pitched roofs of profiled sheet — 10
Flat roofs with sandwich warm deck — 13
Flat roofs with inverted warm deck — 15
Flat roofs with cold deck — 17
Quality control checks for roofs — 19

3 Walls
Rain penetration through cavity and solid walls — 21
Masonry cavity walls — 25
Masonry walls with internal insulation — 27
Masonry walls with external insulation — 30
Walls of timber framed construction — 32
Quality control checks for walls — 34

4 Windows
Window to wall junctions — 35
Double glazed windows — 39
Quality control checks for windows — 42

5 Floors
Concrete ground floors insulated below the structure — 43
Concrete ground floors insulated above the structure — 47
Concrete ground floors insulated at the edge — 52
Suspended timber ground floors — 55
Concrete and timber upper floors — 58
Quality control checks for floors — 61

Appendices
A Sizes for cables enclosed within thermal insulation — 63
B Insulation thickness for water supply pipes — 64
C Sources of information — 65

Index

Thermal insulation: avoiding risks

1 Whole building

SCOPE OF THIS SECTION

This section deals with the building as a whole rather than with individual elements such as roofs, walls, etc.

Heating, ventilation, solar gain, building use and form need to be considered in conjunction with thermal insulation to obtain required internal conditions and optimum energy efficiency.

 This guidance is of particular relevance when a building is being renovated or converted.

PRINCIPAL TECHNICAL RISK

1.1 Inadequate internal conditions and inefficient energy use

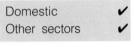

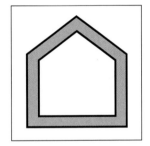

RISK AND AVOIDING ACTIONS

1.1 Inadequate internal conditions and inefficient energy use

Thermal insulation, heating and ventilation should be considered as part of a total design, which takes into account all heat gains and losses. Failure to do so can lead to inadequate internal conditions, eg condensation and mould, and the inefficient use of energy due to overheating or the use of artificial cooling.

DOMESTIC BUILDINGS

1.1 (a) Consider thermal insulation, heating and ventilation together since they interact in providing satisfactory internal conditions (Figure **1**).

(b) Insulate the structure uniformly, avoiding thermal bridging (Figure **1**). Position the insulating layer so as to provide a thermal capacity appropriate to the heating system, eg:

- internal insulation and timber frame construction are compatible with rapid response heating systems, such as warm air,
- external insulation or cavity fill is suited to hot water central heating systems that provide background heat, and for buildings that can exploit solar gains.

(c) Provide a well controlled heating system, installing heat emitters in rooms where heat will not be gained from heated spaces elsewhere in the dwelling, eg in flats and bungalows.

(d) Prevent the distribution of moisture laden air throughout the dwelling, particularly to unheated spaces, by passive or mechanical extract ventilation close to the source.

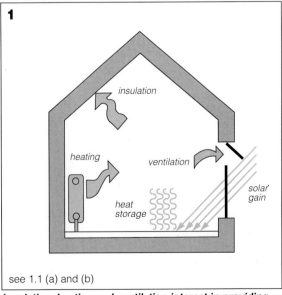

see 1.1 (a) and (b)

Insulation, heating and ventilation interact in providing satisfactory internal conditions

Thermal insulation: avoiding risks

Whole building

NON-DOMESTIC BUILDINGS

(e) Assess risks attached to construction and thermal insulation measures against environmental conditions arising out of the proposed use and life expectancy of the building.

(f) Design to achieve satisfactory internal conditions with the least use of energy by balancing heat losses (fabric and ventilation) and heat gains (from solar radiation, processes, people, lighting and equipment) (Figure **2**).

(g) Choose an orientation and a system of glazing which maximises solar gain in winter and limits heat gain in summer (Figure **2**).

(h) Use thermal capacity to limit the effect of heat gains on internal temperatures when the building is occupied.

(i) Use low energy lighting to limit internal heat gains.

(j) Use controls for heating and lighting to provide satisfactory environmental conditions only when and where necessary.

(k) When natural ventilation is not possible, choose a mechanical ventilation system that can take care of occasional overheating without the need for artificial cooling.

(l) Provide extract ventilation from processes that produce water vapour to reduce the need for elaborate precautions for reducing the risk of condensation within insulated elements.

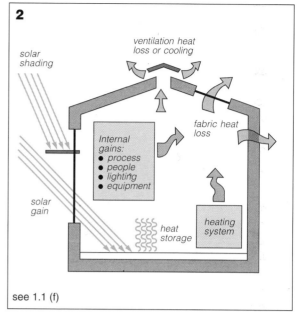

see 1.1 (f)

Heat gains and losses balanced to achieve satisfactory internal conditions

2 Roofs

Pitched roofs with ventilated roof spaces	3
Pitched roofs with sarking insulation	8
Pitched roofs of profiled sheet	10
Flat roofs with sandwich warm deck	13
Flat roofs with inverted warm deck	15
Flat roofs with cold deck	17
Quality control checks for roofs	19

Pitched roofs with ventilated roof spaces

Domestic ✔
Other sectors ✔

CHARACTERISTICS OF THE CONSTRUCTION

The insulation is laid at ceiling level for conventional lofts and in addition at rafter level for room-in-the-roof designs, creating cold roof spaces through which ventilation is essential.

In both cases, it is more efficient thermally to lay the insulation in two layers – the first between, and the second across any areas of thermal bridging, eg ceiling joists and rafters behind inclined ceilings.

 The insulation method is used in renovation work:
- when adding insulation to an accessible loft,
- when converting a loft to a living space,
- when a pitched roof is added above an existing flat roof.

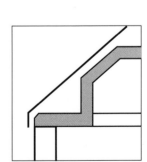

ASSOCIATED TECHNICAL RISKS

2.1 Condensation within roof spaces
2.2 Condensation on the inside of extract ducts passing through unheated roof spaces
2.3 Condensation at thermal bridges
2.4 Risks associated with electrics
2.5 Freezing and contamination of water in pipes and cisterns
2.6 Increased heat loss when insulation is not full depth

Thermal insulation: avoiding risks

Roofs
Pitched roofs with ventilated roof spaces

RISKS AND AVOIDING ACTIONS
2.1 Condensation within roof spaces

Low winter temperatures will occur in roof spaces which are on the cold side of ceiling insulation. If moist air passes through the ceiling, condensation may form on cold surfaces in the roof space.

2.1 (a) Ventilate the roof space to the outside air via openings at the eaves fascia or soffit with a ventilation area equivalent to a 25 mm continuous gap when the pitch is below 15°; a 10 mm gap when the pitch is 15° and above (see BS 5250), (Figure **3**).

(b) Place a 3 to 4 mm mesh across ventilation openings or use a proprietary unit, to prevent the entry of insects, but ensure that the ventilation area requirements are still met (Figure **3**).

(c) Provide an air path to the roof space from the eaves ventilation opening equivalent to a 25 mm continuous gap, irrespective of roof pitch (Figure **3**).

(d) Locate flue outlets more than 850 mm below perimeter vents, or plastics gutters.

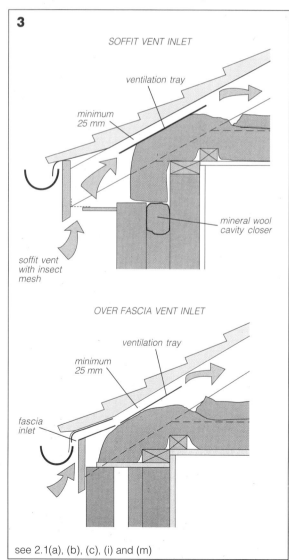

see 2.1(a), (b), (c), (i) and (m)

Ventilation to a cold pitched roof at the eaves

2.1 (e) Provide additional ventilation openings equivalent to a 5 mm continuous gap at the ridge to cross ventilated roofs (Figure **4**):
- when the span is more than 10 metres,
- when the pitch is more than 35°.

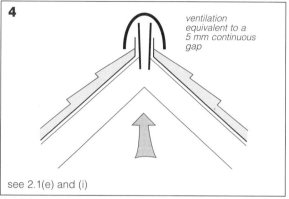

see 2.1(e) and (i)

Additional ventilation to wide or steep roofs

(f) Provide ventilation openings equivalent to a 5 mm continuous gap at high level to roof spaces which are not cross ventilated eaves to eaves, eg mono-pitched roofs (Figure **5**).

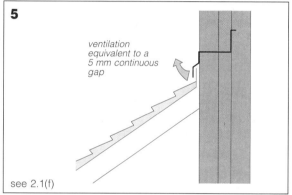

see 2.1(f)

High level ventilation for mono-pitched roofs

(g) Fill gaps in the ceiling at electrical fittings and where pipes rise from high humidity areas, eg bathrooms and wet areas in non-domestic buildings (Figure **6**). Allow for thermal movement of stack vents.

Consider terminating stack vents below the ceiling with an air admittance valve to avoid a hole in the ceiling.

(h) Provide a draughtseal to the loft hatch and bolts or catches to ensure it is compressed (Figure **6**).

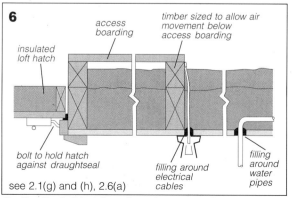

see 2.1(g) and (h), 2.6(a)

Sealing the loft hatch and service penetrations

Thermal insulation: avoiding risks

ROOM-IN-THE-ROOF

2.1 (i) Ventilate roof spaces to the outside air with openings in the eaves fascia or soffit equivalent to a 25 mm continuous gap and additional openings at the ridge equivalent to a 5 mm continuous gap (Figures **3** and **4**).

(j) Provide ventilation to the inclined ceiling of a room-in-the-roof by means of an air path at least 50 mm deep between the insulation and the tiling underlay (Figure **7**). Board insulation is preferable to quilt. Pay particular attention to ventilation paths where the roof line is interrupted by dormers, roof lights, etc.

(k) Provide a vapour control layer, eg 500 gauge (0.12 mm) polyethylene and a breather membrane as a tiling underlay to inclined ceilings of rooms-in-the-roof and parts of the roof where the ventilation path is restricted, eg where there are complicated roof shapes. Lap and seal joints.

The insulated wall to the lower triangle should not be punctured. Storage cupboards should be built within the insulated enclosure (Figure **7**).

(l) Avoid puncturing the vapour control layer (eg for services). Batten out to create a services zone, or route services within a continuous layer of insulation on the warm side of the vapour control layer (Figure **7**). Ensure that the thermal resistance of the construction on the cold side of the vapour control layer is at least three times greater than the resistance of the construction on the warm side.

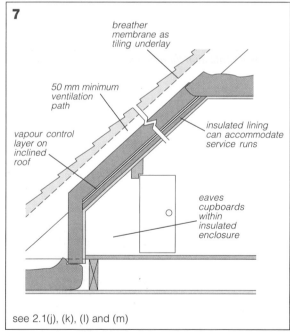

see 2.1(j), (k), (l) and (m)

Room-in-the-roof with eaves cupboards

R 2.1(m) Follow the advice in 2.1(a) to 2.1(l) when insulating an existing roof in a way which requires roof spaces to be ventilated.

When necessary, use over fascia ventilators and ventilating tiles to achieve roof space ventilation (Figure **3**).

2.2 Condensation on the inside of extract ducts passing through unheated roof spaces

The warm moist air extracted from kitchens and bathrooms cools when it passes through ducts in unheated roof spaces. If the air is cooled below its dew point, moisture will condense on the duct wall and drip back into the fan unit.

2.2 (a) Insulate ventilation ducts that pass through unheated roof spaces using a material that gives a thermal resistance of at least 0.6 m^2K/W (Figures **8** and **9**).

(b) Arrange for the safe dispersal of any condensate which forms within the duct.
If the extract discharge is at the eaves, ensure that the duct is laid to a fall. If the extract duct is vertical, provide a condensation trap and a small diameter pipe to drain the condensate to the eaves (Figures **8** and **9**).

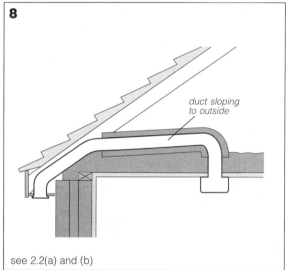

see 2.2(a) and (b)

Insulating a horizontal extract duct

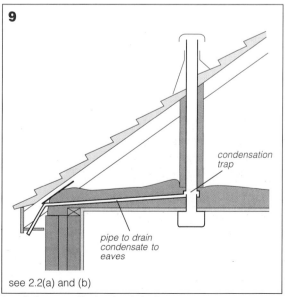

see 2.2(a) and (b)

Insulating a vertical extract duct

Roofs
Pitched roofs with ventilated roof spaces

2.3 Condensation at thermal bridges

Where gaps occur in the insulation, a thermal bridge is created and there is a risk of condensation. A thermal bridge can occur at the junction of a roof or a ceiling with a masonry wall.

2.3 (a) Carry the loft insulation over the wall plate to link with the wall insulation (Figure **10**). Extending the roof truss setting out point beyond the wall plate avoids compressing insulation between the wall plate and the ventilation tray.

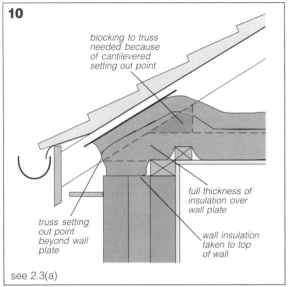

see 2.3(a)

Extending truss setting out point to accommodate full insulation thickness over wall plate

(b) Where it is necessary to close the wall cavity:
- use an insulating block as a closer, but only if no significant differential movement between inner and outer leaves is anticipated, OR
- use mineral wool in a polyethylene cover as a cavity closer to avoid any damage from differential movement between, for example, block inner and brick outer leaves, OR
- use a thin board, eg calcium silicate, as a cavity closer, provided it is fixed in a way that will not result in problems if there is differential movement between inner and outer leaves (Figure **11**).

Do not use dense masonry as a cavity closer.

Where cavity closers are required to perform as cavity barriers, they should provide 30 minutes fire resistance.

(c) At gable walls:
- insulate the gap between the last ceiling joist and the gable wall,
- take both loft and wall insulation at gables to at least 225 mm above ceiling level and use an insulating block for the inner leaf (Figure **12**).

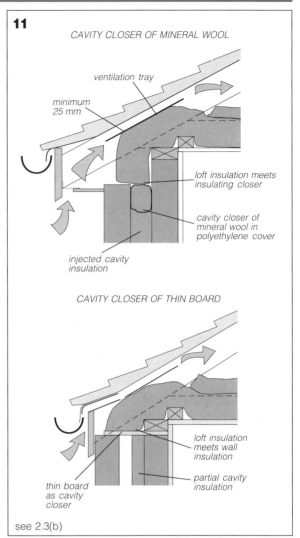

see 2.3(b)

Examples of avoiding thermal bridging at the eaves

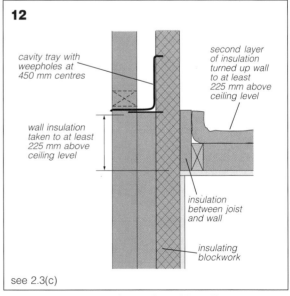

see 2.3(c)

Avoiding thermal bridging at the gable wall

 2.3(d) Add thermal insulation as necessary at the roof perimeter to avoid thermal bridging when adding a pitched roof to an existing flat roof building. Where appropriate, follow the advice in 2.3(a) to 2.3(c).

Pitched roofs with ventilated roof spaces   **Roofs**

2.4 Risks associated with electrics

Electrical cables give off heat when in use. Where cables are covered by thermal insulation they may overheat, increasing the risk of short circuit, or fire starting in combustible loose fill and plastics insulation.

PVC sheathing to cables can have reduced life expectancy if in direct contact with expanded polystyrene insulants.

High wattage recessed light fittings need to be cooled by a through flow of air from the building. This air flow can cause condensation problems in the loft.

2.4 (a) Fix cables to the structure above the insulation so that they can dissipate heat. Cables that pass through, or are enclosed by, insulation may need to be increased in size, even when in conduit. The circuits most likely to be affected are radial circuits serving cookers, immersion heaters, shower units and socket outlets (see Appendix A).

(b) Ensure PVC insulated cables are not in direct contact with expanded polystyrene insulation. Alternatively, run them in conduit.

(c) Where recessed fittings are to be used, specify those designed for compact fluorescent or low voltage tungsten halogen lamps, and locate them within an enclosure, between the joists to dissipate heat. Seal the enclosure to the roof space, but ventilate if necessary to the room.

2.5 Freezing and contamination of water in pipes and cisterns

The better insulated the ceiling, the colder the air temperature in the roof space. Water in pipes and cisterns outside the insulated enclosure may freeze and burst pipes if not drawn frequently during cold weather. Subsequent thawing will cause damage to ceilings and furnishings. Condensation on pipes or cisterns can damage cistern supports and stain ceilings.

2.5 (a) Locate pipes in heated spaces whenever possible, eg below loft insulation or below ceilings. Avoid running hot and cold water pipes closely together, since this will warm the cold supply too much.

(b) Specify pipe insulation for all pipes, including warning pipes, in unheated roof spaces (Figure **13**). The minimum thickness depends on the thermal conductivity of the insulation material (see Appendix B). Where no heat is available from the occupied space, or the building remains unoccupied for long periods, install thermostatically controlled trace heaters for pipes and low output heaters for cisterns.

(c) Insulate the top and sides of cold water cisterns in the roof space. Omit insulation from directly under the cistern to allow warmth from below to reach the cistern base. Insulate the gap below the cistern base by turning up the loft insulation against the tank insulation.

Include the rising main within the insulated enclosure for the tank and insulate it using a closed cell vapour resistant material with all joints covered with vapour resistant tape, or enclose fibrous insulation within a vapour control layer.

Use a moisture resistant board, eg Type C4 chipboard to BS 5669, as the cistern support to avoid damage from any condensation which may form on the outside of the cistern (Figure **13**).

(d) Provide a cistern with a closely fitting cover, a screened air inlet and a screened warning pipe to reduce contamination of stored water by insects. Install an isolating valve on the supply to the cistern to allow for servicing and cleaning (Figure **13**).

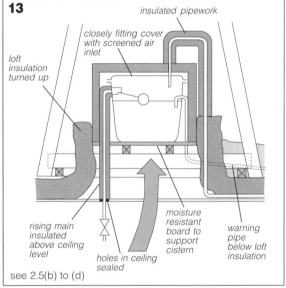

Insulated pipes and cisterns in roof space

2.6 Increased heat loss where insulation is not full depth

Loft insulation may need to be deeper than the ceiling joists and, if they are obscured, there is a risk of falling through the ceiling. Access boarding to water services or to storage may limit insulation thickness.

2.6 (a) To reduce condensation on the underside of access boarding, use:
- slatted boarding, OR
- support panel flooring at least 25 mm above the top of the insulation to allow air movement between the flooring and the insulation (Figure **6**).

(b) Insulate the loft hatch to the highest practical level, or choose a proprietary unit with the best available insulation properties.

Roofs

Pitched roofs with sarking insulation

Domestic ✓
Other sectors ✓

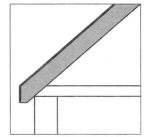

CHARACTERISTICS OF THE CONSTRUCTION

The insulation is placed above and between the rafters creating a warm structure and loft space which is particularly suitable for room-in-the-roof construction. This sarking insulation is normally low vapour permeability rigid plastics boards.

 The insulation method is relevant if the roof covering is to be replaced or an existing roof is to be converted to living space. It is a particularly useful alternative where it is difficult to provide eaves ventilation to a cold pitched roof.

ASSOCIATED TECHNICAL RISKS

2.7 Increased heat loss due to air movement
2.8 Condensation within the construction
2.9 Condensation at thermal bridges
2.10 Condensation on pipes and ducts

RISKS AND AVOIDING ACTIONS

2.7 Increased heat loss due to air movement

Gaps in the insulation layer allow air to move between the warm interior and the cold exterior, causing increased heat loss.

2.7 (a) Fill all gaps between insulation boards and with the structure at gable ends, ridge, eaves, abutments and roof penetrations with flexible sealant or expanding foam (Figures **14** to **16**).

(b) Ensure a tight fit between roof boards and between roof boards and rafters.
This can be achieved by using proprietary rebated insulation boards (sized to suit standard rafter spacings) as templates when positioning rafters (Figures **15** and **16**). Their use also avoids the rafters being forced out of alignment if they are not spaced accurately.

2.8 Condensation within the construction

Water vapour in the internal air can condense on cold surfaces within the construction if it penetrates the insulation layer and cannot permeate through the tiling underlay.

2.8 (a) Tightly fit the insulation layer as in 2.7(b) and ensure that a breather membrane is used as the tiling underlay (Figure **15**), or ventilate beneath it.

(b) Ensure that the tiling underlay is draped across pressure impregnated counter battens to avoid it touching the insulation, as some types of underlay allow liquid water to pass through when in contact with insulation (Figure **15**).

 (c) Remove any existing loft insulation at ceiling level so that insulation is concentrated at rafter level.

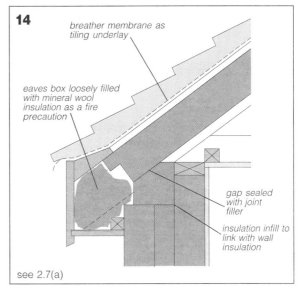

Warm pitched roof at the eaves

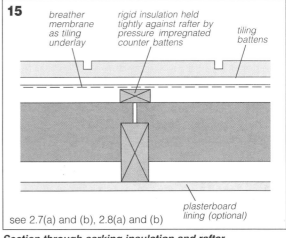

Section through sarking insulation and rafter

Thermal insulation: avoiding risks

2.9 Condensation at thermal bridges

Gaps or lack of continuity in the insulation create thermal bridges allowing condensation to form on cold surfaces.

(a) Ensure continuity between roof and wall insulation at eaves and verge. Insulate gable walls to link with the roof insulation at the verge (Figure **16**).

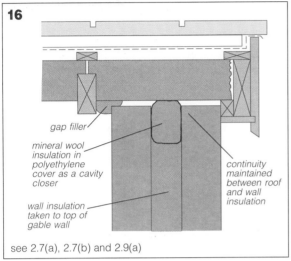

see 2.7(a), 2.7(b) and 2.9(a)

Junction of sarking insulation and gable wall

2.10 Condensation on pipes and ducts

Moisture from air in the warm interior can condense on the cold surfaces of pipes and ducts which pass through the roof construction and are exposed within the insulated space. The condensate can build up to an extent where it may damage the structure and finishes.

2.10 (a) Insulate pipes and ducts exposed to moist air if it is considered that condensate will cause damage.

(b) Use a closed cell vapour resistant insulant and wrap joints with vapour resistant tape, or enclose fibrous insulation within a vapour control layer, to restrict water vapour from reaching the pipe.

Roofs

Pitched roofs of profiled sheet

Domestic ✗
Other sectors ✓

CHARACTERISTICS OF THE CONSTRUCTION

There are two insulation concepts for profiled sheet roofing:

- WARM ROOF in which insulation is sandwiched between, and in full contact with, two sheets which are fixed above purlins. Composite panels can be formed by foaming between the sheets, or the construction can be built up in layers using specially shaped insulants to match the sheet profiles. The system uses vapour resistant insulants and, if correctly sealed at the joints, needs no separate vapour control layer or breather membrane.
- COLD ROOF in which insulation is usually above purlins with a ventilation space between the outer sheet and the insulation. Vapour permeable insulation can be used but, since it is necessary to control condensation, a vapour control layer and breather membrane are required.

R This type of construction is appropriate when a similar, poorly insulated cladding is at the end of its useful life, or when roofing over existing flat roofs.

ASSOCIATED TECHNICAL RISKS

2.11 **Condensation within the construction**
2.12 **Condensation at thermal bridges**
2.13 **Condensation on internal rainwater downpipes**
2.14 **Corrosion of sheeting and fixings**
2.15 **Delamination of composite panels**

Note: In assessing these risks, consideration should be given to the expected life and use of the building, as well as to the anticipated internal and external environmental conditions. Guidance on the suitability of construction methods for different building types is given in BS 5427 and in publications by British Steel.

RISKS AND AVOIDING ACTIONS

2.11 Condensation within the construction

This is a substantial risk with cold roofs. Condensate can form on the inner face of the outer sheet, due to radiation to the night sky, and sometimes freezes there. Condensate can saturate the insulation and corrode fixings and the edges of metal sheet. Condensate may eventually drip from purlins or sheet fixings onto ceilings and into the building.

WARM ROOF
2.11 (a) Use a warm roof if possible and ensure that:

- all voids are filled with foamed plastics insulation
- joints incorporate a vapour seal to prevent moist air reaching the cold overlapping metal sheet (Figures **17** and **18**).

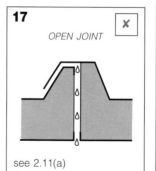

17 ✗ OPEN JOINT
see 2.11(a)

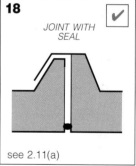

18 ✓ JOINT WITH SEAL
see 2.11(a)

COLD ROOF
(b) Use cold roofs only where it is possible to:

- provide extract ventilation close to processes producing water vapour,
- incorporate a vapour control layer on the warm side of the insulation,
- detail the roof so that condensate drains to the eaves gutter by means of a breather membrane which allows water vapour to pass through from the inside but allows liquid water to drain,
- ensure adequate ventilation through rib voids above the insulation from eaves to ridge.

Pitched roofs of profiled sheet   **Roofs**

2.11 (c) Choose, for preference, a system in which the insulation and inner support panels are above the purlins. If purlins need to be concealed, add a separate decorative lining in addition to the insulated construction (Figure **19**).

(d) Form a vapour resistant inner layer by:
- laying a vapour control layer of 500 gauge (0.12 mm) polyethylene over the support panel and sealing the joints with vapour resistant tape (Figure **19**), OR
- sealing all joints between support panels.

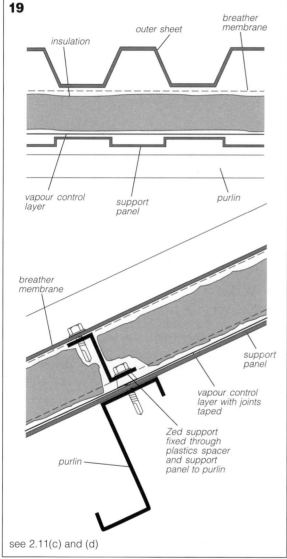

see 2.11(c) and (d)

Traditional design with Zed support and plastics spacers – COLD ROOF

(e) Ensure that all fixings to the purlin incorporate a method of sealing the holes against the passage of water vapour (Figure **21**).

(f) Form a vapour tight seal where the vapour control layer is punctured by pipes or ducts.

(g) Cut down or perforate filler strips at eaves and ridge to allow ventilation of rib voids and release of condensate (Figure **22**).

(h) Avoid end laps whenever possible. When this is unavoidable, seal joints at inside as well as outside laps (Figure **20**). This minimises the amount of condensate that can collect between the sheets and attack corrodible fixings. It is unnecessary to seal the end laps when the pitch is greater than 15° and the lap is at least 200 mm, on sites with sheltered exposure to driving rain.

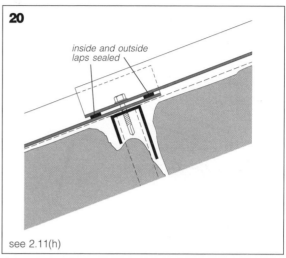

see 2.11(h)

Sealing at end laps – COLD ROOF

2.12 Condensation at thermal bridges

Purlins and spacer sections commonly link inner and outer metal components. Internal gutters are often uninsulated. Some types of rooflight break the continuity of the construction.

COLD ROOF

2.12 (a) Provide extract ventilation close to processes producing water vapour.

(b) Create a thermal break by inserting a non-metallic spacer at some point between the outer profiled sheet and the purlin (Figure **21**).

(c) Ensure that insulation is packed beneath support bars.

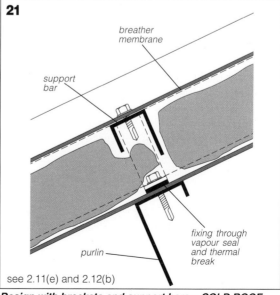

see 2.11(e) and 2.12(b)

Design with brackets and support bars – COLD ROOF

Thermal insulation: avoiding risks   11

2.12 (d) Design the roof to fall to external eaves gutters (Figure **22**). If internal gutters are required, specify that they are insulated with closed cell vapour resistant insulation that restricts water vapour from reaching the cold metal surface. Joints between sections of insulation should be sealed with vapour resistant tape.

(e) Whenever possible, specify site-assembled double skin rooflights that provide continuity with the lining sheets.

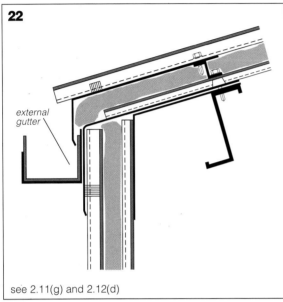

see 2.11(g) and 2.12(d)
External eaves gutter – COLD ROOF

2.13 Condensation on internal rainwater downpipes

The walls of rainwater downpipes and other pipes that penetrate the roof (ie stack vents) present a cold surface on which moisture from the internal air can condense. This can run down and damage internal finishes.

WARM AND COLD ROOFS
2.13 (a) Insulate internal rainwater downpipes and other pipes that penetrate the roof if they pass through spaces with a high humidity and if any condensate will damage the structure or internal finishes.

Use a closed cell vapour resistant insulant and wrap joints with vapour resistant tape or enclose fibrous insulation within a vapour control layer to restrict water vapour from reaching the pipe.

2.14 Corrosion of sheeting and fixings

The coating on the underside of the sheeting is often less corrosion-resistant than the external coating, making the underside susceptible to corrosion due to condensation. Moisture between the sheets can accelerate corrosion of fixings.

WARM AND COLD ROOFS
2.14 (a) Choose an external sheeting with a coating, on its under surface, which is capable of resisting corrosion caused by condensation.

(b) Choose fixings of stainless steel for buildings. Where the relative humidity is not likely to rise above 65%, or for buildings designed for a short lifetime, carbon steel fixings with a protective coating are adequate.

2.15 Delamination of composite panels

With composite panels of warm roof construction, an inadequate bond during manufacture between the insulation and the outer sheet may be broken due to extreme thermal movement of the outer sheet. This results in bowing and possibly condensation in the newly formed voids.

WARM ROOF
2.15 (a) Seek third party assurance from the manufacturer that there is an adequate bond between the upper sheet and the insulation.

(b) Choose light rather than dark colours whenever possible to minimise thermal bowing.

Roofs

Flat roofs with sandwich warm deck

Domestic ✔
Other sectors ✔

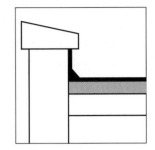

CHARACTERISTICS OF THE CONSTRUCTION

The insulation is placed above the roof deck, but below the weatherproof membrane. Normally, there should be no insulation below the deck. Ventilation is not required, since the insulation is bedded on a continuous vapour control layer. The method is used above timber, concrete and metal structural decks.

R This insulation method is appropriate when the weatherproof membrane of a concrete or timber warm deck flat roof has failed and needs to be renewed or where the membrane is in good condition, but the insulation needs to be upgraded. For timber flat roofs, the method may also be used to replace a cold deck construction where only the structural joists are in a satisfactory condition.

ASSOCIATED TECHNICAL RISKS

2.16 Fatigue of weatherproof membrane
2.17 Insulant damaged by heat during construction
2.18 Fire spread from a melting or burning weatherproof membrane and combustible insulation
2.19 Condensation within the construction
2.20 Condensation at thermal bridges

RISKS AND AVOIDING ACTIONS

2.16 Fatigue of weatherproof membrane

Placing insulation directly below the weatherproof membrane will increase the membrane's surface temperature fluctuations. On sunny days the insulation slows the heat flow to the deck causing a build up in the membrane's surface temperature. At night the membrane has a lower surface temperature as it loses heat by radiation to the night sky.

2.16 (a) For built-up roofing, use Class 5 high tensile membranes to BS 747 or newer high performance membranes that can accommodate stresses and remain flexible over a wide temperature range. Partial bonding of the first layer can help distribute any stresses evenly beneath the weatherproof membrane.

(b) Protect weatherproof membranes with a solar reflective finish to reduce thermal stress.

(c) For polymeric single ply materials, follow manufacturers' recommendations on bonding specification.

(d) Provide sufficient falls for membranes of all kinds to ensure drainage and avoid ponding, and to reduce stresses on the membrane during heating and cooling.

2.17 Insulant damaged by heat during construction

Some insulation materials can be damaged by high temperatures when laying asphalt or using hot bitumen to bond built-up felt roofing.

2.17 (a) Avoid damage to expanded polystyrene insulation by:
- protecting the insulation with a layer of corkboard or bitumen-impregnated fibreboard, OR
- using a composite insulation board with integral heat protection.

2.18 Fire spread from a melting or burning weatherproof membrane and combustible insulation

A fire source inside the building may cause cellular plastics insulants and weatherproof membranes laid directly onto unprotected metal decks to melt and/or decompose. The flammable gases generated and any molten materials, eg bitumen or polystyrene, may penetrate joints in the roof deck and increase the spread of fire below.

2.18 (a) Choose a non-combustible insulant or place a gypsum board between the metal deck and cellular plastics insulation. Alternative methods of improved fire protection for insulated roof decks are available and could be usefully discussed with a fire expert.

Thermal insulation: avoiding risks

Roofs
Flat roofs with sandwich warm deck

2.19 Condensation within the construction

If water vapour is able to penetrate the insulation, it can condense on the cold weatherproof membrane and cause degradation of the insulation and bubbling beneath the weatherproof membrane, which in time will lead to its breakdown.

2.19 (a) Lay a vapour control layer of high performance roofing felt to BS 747 immediately below the insulation, fully supported by the roof deck. The vapour control layer should be fully bonded to the deck and have all laps sealed with hot bitumen.

At the perimeter, the vapour control layer should be turned back 150 mm over the insulation and bonded to the weatherproof membrane. Where insulation is compressible, it should have a dense topping to support the weatherproof membrane if roof access is anticipated (Figures **23** and **24**).

(b) In timber construction, restrict the passage of moist air from the structural roof void to the eaves box by installing a vertical strip of closed cell rigid insulation and sealing all gaps, eg with expanding foam, to minimise condensation on cold surfaces within the eaves box.

R (c) Ensure that the greater thermal resistance is on the cold side of the deck when it is not possible to avoid having some insulation below the deck. The thermal resistance of the construction on the cold side of the deck should be at least three times greater than the resistance of the construction below the deck.

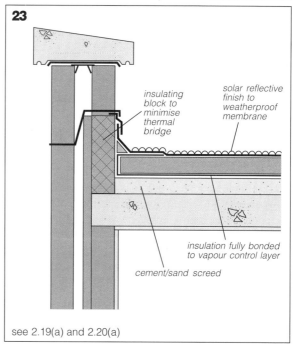

see 2.19(a) and 2.20(a)

Sandwich warm deck roofing above a concrete structure

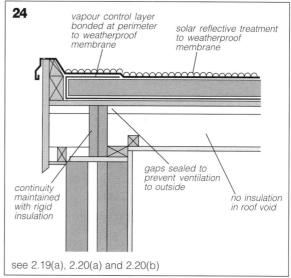

see 2.19(a), 2.20(a) and 2.20(b)

Sandwich warm deck roofing above a timber structure

2.20 Condensation at thermal bridges

Where the continuity of the insulating layer is broken, particularly at the roof/wall junction and where pipes penetrate the roof, a thermal bridge occurs and there is the risk of surface condensation.

2.20 (a) To avoid a thermal bridge at the roof/wall junction and at changes in level:
- maintain continuity of insulation, OR
- overlap roof and wall insulation and introduce insulation material between (Figures **23** and **24**).

When the insulation is above a concrete deck, breaks in the insulation at upstands and plinths within the main roof area will not create a serious risk of surface condensation.

(b) In timber construction, insulate the perimeter of the structural roof void to maintain continuity between roof and wall insulation (Figure **24**).

(c) Insulate internal rainwater downpipes and other pipes that penetrate the roof if they pass through spaces with a high humidity and if any condensate will damage the structure or internal finishes.

Use a closed cell vapour resistant insulant and wrap joints with vapour resistant tape or enclose fibrous insulation within a vapour control layer to restrict water vapour from reaching the pipe.

Flat roofs with inverted warm deck

Domestic ✔
Other sectors ✔

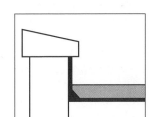

CHARACTERISTICS OF THE CONSTRUCTION

The insulation is placed above the weatherproof membrane. It is protected from UV degradation and held down against wind uplift by a ballast layer and edge restraint. A filter layer is laid immediately below the ballast. The method can be used above concrete, metal and timber structural decks.

Concrete decks are capable of supporting a ballast layer of paving slabs or pebbles.

For lightweight decks that cannot carry heavy ballast, use a permeable sheet topping to restrain tongued and grooved insulation boards to create a fully interlocking layer. Care needs to be taken with timber decks to ensure that the structure is capable of supporting the inverted roof system.

For new construction, there should be no materials with a high thermal insulation value below the deck, except where insulation is needed to reduce condensation on the underside of lightweight metal decks.

R This method may be used when the previous construction is to be renewed or when it is sufficient to upgrade an existing poorly insulated construction.

When upgrading an existing roof, it will be necessary to check the thermal resistance of any insulation previously installed below the weatherproof membrane and the ability of the roof to support the increased load.

ASSOCIATED TECHNICAL RISKS

2.21 Increased heat loss due to rain cooling
2.22 Condensation through chilling of lightweight metal decking
2.23 Condensation within the construction
2.24 Condensation at thermal bridges
2.25 Degradation of the insulant
2.26 Abrasion of the weatherproof membrane

RISKS AND AVOIDING ACTIONS

2.21 Increased heat loss due to rain cooling

The weatherproof membrane of an inverted roof remains close to room temperature at all times of the year. However, cold rainwater which percolates through the ballast and joints in the insulation will flow over this relatively warm surface, causing increased heat loss.

2.21 (a) Add 20% to the insulation thickness required to meet building regulation requirements to offset the intermittent cooling effect of rain.

2.22 Condensation through chilling of lightweight metal decking

There is a risk of localised condensation on the underside of lightweight decks if melting snow or cold rainwater percolates through the insulation layer and chills the deck. The risk is greatest where the internal temperature and relative humidity are high. Concrete roof decks, because of their high thermal mass, are less affected by chilling and do not normally suffer from surface condensation.

2.22 (a) Ensure that the thermal resistance of the construction between the weatherproof membrane and the heated space is at least 0.15 m²K/W, but not more than 25% of the thermal resistance of the whole construction.

(b) Restrict the amount of rainwater able to reach the weatherproof membrane by tightly butting insulation boards and trimming them tightly around projections and at upstands. The use of insulation boards with an overlapping edge profile is advantageous.

Thermal insulation: avoiding risks

Roofs
Flat roofs with inverted warm deck

2.23 Condensation within the construction

> *Where there is too high a proportion of the total thermal resistance below the weatherproof membrane, there is a risk of condensation within the construction.*

R **2.23(a)** When some insulation is needed below the deck or previously installed insulation remains in the construction, ensure that the greater thermal resistance of the construction is above the weatherproof membrane. A ratio of 3:1, above to below, is recommended.

2.24 Condensation at thermal bridges

> *Where the continuity of the insulating layer is broken, particularly at the roof/wall junction and where pipes penetrate the roof, a thermal bridge occurs and there is a risk of surface condensation.*

2.24(a) Overlap wall and roof insulation to extend the thermal bridge path, if necessary adding thermal insulation to edge beams to achieve continuity with external insulation (Figures **25** and **26**).

(b) Insulate parapet upstands or use an insulating block in the thermal bridge path between the wall and roof insulation (Figures **27** and **28**).

(c) For insulating rain water pipes, see 2.20(c).

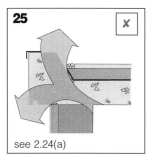
see 2.24(a)

see 2.24(a)

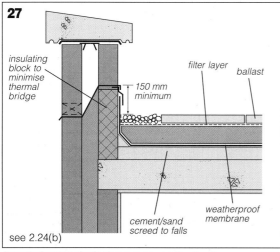
see 2.24(b)

Inverted warm deck above a concrete structure

2.25 Degradation of the insulant

> *Being above the weatherproof membrane, the insulant is subject to degradation from freeze/thaw cycles, wetting and UV light if ballast moves. The insulation can also be damaged by chipping on existing roof coverings.*

2.25(a) Use an insulant which has low water absorption and is frost resistant.

(b) Ensure that plastics foam insulants are not exposed to UV light. Use flashings to protect the insulant at upstands.

Provided the ballast layer is continuous, it will protect the insulant from UV degradation.

Follow the calculation in BRE Digest 311 to assess the risk of wind scouring gravel ballast and exposing the insulant. If necessary, increase the gravel size, use paving slabs in vulnerable areas, or increase the parapet height.

R **(c)** If any chippings remain bonded to an existing weatherproof membrane, lay a cushioning layer of polyethylene foam below the insulation.

2.26 Abrasion of the weatherproof membrane

> *Grit washed down between the insulation boards can cause abrasion and eventually puncture weatherproof membranes, particularly single layer membranes.*

2.26(a) Lay a cushioning layer above single layer membranes. Turn up the cushioning layer at upstands.

(b) Place a suitable geotextile filter membrane above the insulation and below the ballast layer. Turn up the filter membrane at upstands.

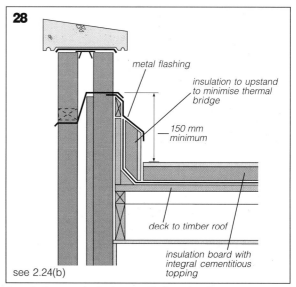
see 2.24(b)

Inverted warm deck above a timber structure

Roofs

Flat roofs with cold deck

Domestic ✔
Other sectors ✔

CHARACTERISTICS OF THE CONSTRUCTION

The insulation is placed at ceiling level with a void ventilated to the outside between the insulation and the deck. It should be considered only for timber construction.

The cold deck roof is considered a poor option in the temperate, humid climate of the UK, where sufficient ventilation may not be achieved in sheltered locations or in windless conditions, even when the roof is correctly designed.

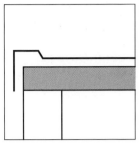

R It is usually not possible to upgrade the thermal insulation value of an existing cold deck roof. The preferred option is to convert it to a sandwich or inverted warm deck flat roof.

ASSOCIATED TECHNICAL RISKS

2.27 Condensation within the roof
2.28 Condensation at thermal bridges

RISKS AND AVOIDING ACTIONS

2.27 Condensation within the roof

If humid air which enters the roof from inside or outside the building is not ventilated away, moisture can condense on the underside of the roof deck and drip into the insulation, damaging ceilings and eventually the timber structure. It can also cause early corrosion of metal roof coverings. The risk of condensation is high.

2.27 (a) Restrict humid air from entering the roof space by providing a continuous vapour control layer of 500 gauge (0.12 mm) polyethylene across the roof area, with joints lapped and taped, sealed to the wall at the edges.

To avoid puncturing the vapour control layer:
- avoid running services within the roof structure, OR
- form a services zone immediately above the ceiling (Figure **30**).

(b) Where services pass vertically through the roof (eg stack vents), seal the joint between the pipe and the ceiling (allowing for thermal movement) and seal the vapour control layer to the pipe.

Alternatively, avoid taking services through the roof structure. In the case of stack vents, terminate the pipe, where possible, with an air admittance valve and locate it in a ventilated duct to avoid damage.

(c) Do not use a cold deck solution if cross ventilation is restricted or blocked, eg by solid strutting or cavity barriers, or if the structure spans between parapet or abutment walls.

2.27 (d) Provide cross ventilation to each and every void to the outside air, via an unrestricted air space above the insulation (Figures **29** and **30**).

(e) For spans up to 5 m, provide ventilation openings to each discrete cavity on opposite sides of the roof with a ventilation area equivalent to a continuous 25 mm gap (for spans up to 5 m) or 30 mm (for 5 m to 10 m spans), or in total, 0.6% of the roof plan area, whichever is the greater (Figure **29**).

(f) Provide an unrestricted air space above the insulation of at least 50 mm (for spans up to 5 m) or 60 mm (for 5 m to 10 m spans), (Figure **29**).

(g) Prevent the entry of insects by placing 3 to 4 mm mesh across ventilation apertures (Figure **30**).

(h) Ensure that the mesh is taken into account when calculating the ventilation area.

(i) Do not locate flue outlets less than 850 mm below perimeter vents or plastics gutters.

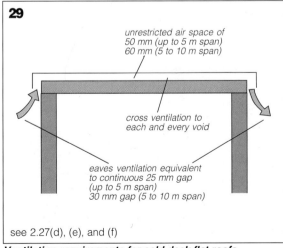

29

unrestricted air space of 50 mm (up to 5 m span) 60 mm (5 to 10 m span)

cross ventilation to each and every void

eaves ventilation equivalent to continuous 25 mm gap (up to 5 m span) 30 mm gap (5 to 10 m span)

see 2.27(d), (e), and (f)

Ventilation requirements for cold deck flat roofs

Thermal insulation: avoiding risks

Roofs — Flat roofs with cold deck

2.28 Condensation at thermal bridges

Where the continuity of the insulating layer is broken, particularly at the roof/wall junction and where pipes penetrate the roof, a thermal bridge occurs and there is the risk of surface condensation.

2.28 (a) Carry the roof insulation over the wall plate to meet the wall insulation (Figure **30**).

(b) Where it is necessary to close the wall cavity:
- use an insulating block as a closer, but only if no significant differential movement between inner and outer leaves is anticipated, OR
- use mineral fibre in a polyethylene cover as a cavity closer to avoid any damage from differential movement between, for example, block inner and brick outer leaves, OR
- use a thin board, eg calcium silicate, as a cavity closer, provided it is fixed in a way that will not result in problems if there is differential movement between inner and outer leaves (Figure **30**).

Do not use dense masonry as a cavity closer.

Where cavity closers are required to perform as cavity barriers, they must comply with building regulations.

(c) Insulate internal rainwater downpipes and other pipes that penetrate the roof if they pass through spaces with a high humidity and if any condensate will damage the structure or internal finishes.

Use a closed cell vapour resistant insulant and wrap joints with vapour resistant tape, or enclose fibrous insulation within a vapour control layer, to restrict water vapour from reaching the pipe.

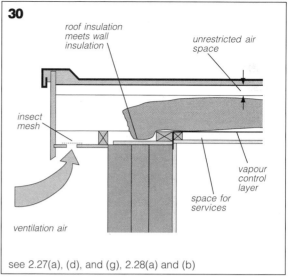

see 2.27(a), (d), and (g), 2.28(a) and (b)

Avoiding vapour transfer to the roof space and thermal bridging at the eaves

Quality control checks for roofs

PITCHED ROOFS WITH VENTILATED ROOF SPACES

- Is the insulation laid without gaps and does it link with the wall insulation?
- Is ventilation provided and are routes clear to the roof space and above any insulation on the slope?
- Are cavity closers of low conductivity material which, when built in, will not cause problems in the event of differential movement between the leaves of the cavity wall?
- Are all holes for services etc through the ceiling or through vapour control layers sealed against air leakage?
- Are cold water cisterns insulated on the top and sides?
- Is all pipework insulated and are thermostatic trace heaters (when installed) functioning properly?
- Are electric cables above the insulation or correctly rated when in contact with or passing through thermal insulation?

PITCHED ROOFS WITH SARKING INSULATION

- Is insulation of low permeability, fitted without gaps and linking with wall insulation?
- Is a breather membrane used as the tiling underlay?

PITCHED ROOFS OF PROFILED SHEET

- Is a continuous vapour control layer with lapped and taped joints fitted above the support panel in cold roofs?
- Is a breather membrane placed above the insulation in cold roofs and does it drain into the gutter?
- Is the ventilation route through rib voids clear at eaves and ridge in cold roofs?
- Are fixings of specified type with all washers, seals etc correctly assembled?

FLAT ROOFS WITH SANDWICH WARM DECK

- Is a continuous vapour control layer of high performance felt bonded to the deck, turned up at the perimeter and all joints sealed?
- Is the insulation fitted without gaps and does it link with the wall insulation?
- Is the insulation being kept dry until roof is sealed?
- Is expanded polystyrene protected from hot bitumen with fibreboard or corkboard?

FLAT ROOFS WITH INVERTED WARM DECK

- Is some insulation placed beneath the weatherproofing membrane where specified for lightweight decks?
- Is drainage from the roof effective?
- Do all insulation boards fit tightly together and is any upstand insulation in position?
- Is the ballast layer being applied soon after fitting the insulation?
- Is a filter membrane laid beneath the ballast layer?
- Is a cushioning layer placed above single layer weatherproof membranes?
- Have steps been taken to avoid wind scouring of pebble ballast?
- Has the insulation been protected from UV light?

FLAT ROOFS WITH COLD DECK

- Is the vapour control layer continuous and fully sealed with no penetrations for services?
- Are ventilation openings and cross ventilation routes clear over the whole roof area?
- Is the insulation fitted without gaps and does it link with the wall insulation?

Thermal insulation: avoiding risks

3 Walls

Rain penetration through cavity and solid walls	21
Masonry cavity walls	25
Masonry walls with internal insulation	27
Masonry walls with external insulation	30
Walls of timber framed construction	32
Quality control checks for walls	34

Rain penetration through cavity and solid walls

CHARACTERISTICS OF THE CONSTRUCTION

This sub-section covers a single technical risk which is common to all insulated masonry walls. The masonry may be solid or cavity construction and of facing masonry or masonry with an additional protective finish such as render or tile hanging. Insulation may be within the cavity, within a timber frame or as internal or external insulation applied to the wall surface.

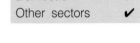

 This guidance is appropriate when adding extensions or when upgrading the thermal performance of existing masonry walls. The condition and history of existing walls can influence the choice of insulation. The incidence of rain penetration problems following cavity fill is about ten times less for existing walls than for new walls.

Note: This technical risk is referred to in all subsequent sub-sections of the Walls section to avoid repeating the complex issues.

RISK AND AVOIDING ACTION

3.1 Rain penetration through masonry walls

Rainwater will penetrate the outer leaves of masonry walls under certain conditions of driving rain. This can lead to dampness reaching the inside causing damage to the wall and decorations and reducing thermal performance.

The performance of the wall in resisting water crossing the cavity is affected by the amount of wind driven rain, the presence of a finish or cladding which protects the wall and the quality of the design, specifications and workmanship.

Moisture transmission to the inner leaf of cavity walls is more likely where walls contain building defects and cavity insulation.

3.1 (a) Follow the procedures below to minimise the risk of rain penetration.

- Follow the calculation or assessment procedures in current British or CEN standards or third party certificates appropriate to the wall construction, location and insulation measure. Ensure that installation is carried out in accordance with relevant standards or certificates by registered persons where appropriate, OR
- Follow the simplified procedure on the following pages for walls up to 12 m high.

 (b) Follow the inspection, assessment and installation procedures in relevant British or CEN standards or third party certificates.

Thermal insulation: avoiding risks

Walls

Rain penetration through cavity and solid walls

Exposure zones	Approximate wind driven rain * [litres/m² per spell]
1	Less than 33
2	33 to less than 56.5
3	56.5 to less than 100
4	100 or more

Key to map

Determining the suitability of proposed wall constructions

- Determine the national exposure zone from the map on this page and apply it to Table 1 opposite. Where necessary modify the zone to allow for the features given below, or local knowledge and practice.

- Add one to the map zone where conditions accentuate wind effects such as on open hillsides or in valleys where the wind is funnelled onto the wall.

- Subtract one from the map zone where walls are well protected by trees or buildings or do not face the prevailing wind.

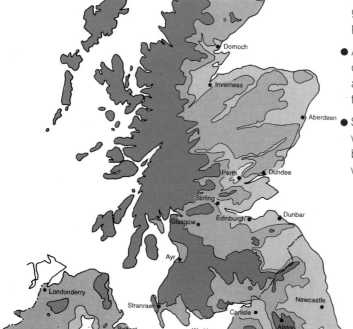

Note:
The wall spell index can be more accurately calculated from the large scale maps and correction factors given in BS 8104, then interpreted using the values in the key to the map above and Table 1 opposite.

* Maximum wall spell index derived from Figure 1 in BS 8104.

22   *Thermal insulation: avoiding risks*

Rain penetration through cavity and solid walls **Walls**

Proposed wall constructions		Maximum recommended exposure zone for each construction						
		Impervious cladding		Rendered finish		Facing masonry		
Construction type and insulation method	Minimum width of filled cavity or clear cavity [mm]	Full height of wall	Above facing masonry*	Full height of wall	Above facing masonry*	Tooled flush joints	Recessed mortar joints	Flush sills and copings
CAVITY FILL INSULATION ● Built-in full fill	filled cavities 50 75 100	4 4 4	3 3 4 (3)	3 4 (3) 4 (3)	3 (2) 3 (2) 3 (2)	2 (1) 3 (2) 3 (2)	1 1 1	1 1 2 (1)
● Injected-fill **not** UF foam	50 75 100	4 4 4	2 3 3	3 4 4	2 (1) 3 (2) 3 (2)	2 (1) 3 (2) 3 (2)	1 1 1	1 1 1
● Injected-fill UF foam	50 75	4 4	2 2	3 3	2 (1) 2 (1)	1 2 (1)	1 1	1 1
PARTIAL FILL ● Residual clear cavity	clear cavity 50	4	4 (3)	4	4 (3)	3 (2)	1	1
INTERNAL INSULATION ● Structural clear cavity (including timber frame) ● Solid masonry wall	clear cavity 50 100 notional cavity	4 4 4	3 4 N/A	4 (3) 4 3	3 (2) 4 N/A	3 (2) 4 N/A	1 2 (1) N/A	1 2 (1) N/A
EXTERNAL INSULATION ● Structural clear cavity ● Solid masonry wall	clear cavity 50 N/A	4 4	N/A N/A	4 3	N/A N/A	N/A N/A	N/A N/A	N/A N/A

* Refer to Figure 31 for an interpretation of walls with impervious cladding or render above facing masonry.

Table 1 – Maximum recommended zones for walls built following the guidance in the notes below. Consider restricting use to zones in brackets if design or construction standards cannot be relied on

Notes to Table 1

Cavities
- Cavities to be not less than the stated width and free of obstructions which will transmit water towards the inner leaf.

Solid masonry
- Walls insulated internally to be at least 328 mm thick if of brickwork, 250 mm if of aggregate blockwork and 215 mm if of autoclaved aerated concrete blockwork with a notional cavity between the masonry and the insulation.

Overhangs
- Sills, copings, string courses and drips below cladding or render to project at least 50 mm and incorporate a throating (Figure **32**). Flush sills and copings give no protection to the wall below.
- Overhangs at eaves and verges to be at least 50 mm and incorporate a throating. The greater the overhang, the greater the protection.

Stop ends
- Watertight stop ends to be secured at the ends of all cavity trays or lintels intended to function as cavity trays to prevent water discharging from the ends into the cavity (Figure **34**).
- For complicated situations, building sequence to be considered and clear drawings, preferably isometric, provided to obtain pre-formed cavity trays and stop end profiles.

Cavity trays
- Cavity trays to be provided:
  – at all interruptions which are likely to direct rainwater across the cavity, such as rectangular ducts, lintels and recessed meter boxes
  – above cavity insulation which is not taken to the top of the wall, unless that area of wall is protected by impervious cladding
  – above lintels in walls in exposure zones 4 and 3 and in zones 2 and 1 where the lintel is not corrosion resistant and intended to function as its own cavity tray
  – continuously above lintels where openings are separated by short piers
  – above openings where the lintel supports a brick soldier course.
- Cavity trays to rise at least 140 mm from the outer to the inner leaf, to be self-supporting or fully supported, and have joints lapped and sealed. The rise across the cavity should be at least 100 mm (Figure **33**).

Weepholes
- Weepholes to be installed at not more than 900 mm centres to drain water from cavity trays and from the concrete cavity infill at ground level. When the wall is to be cavity filled, it is advisable to reduce the spacing.
- At least two weepholes to be provided to drain cavity trays above openings.

- Provide means of restricting the entry of wind driven rain through weepholes in walls in exposure zones 4 and 3, including at ground level.

Mortar and render
- A mortar mix whose strength is compatible with the strength and type of masonry unit must be specified to minimise cracking, especially for concrete and calcium silicate units.
- Tooled mortar joints, either bucket handle or weathered, to be used. Recessed or raked joints to be used only in exposure zone 1 with 50 mm clear cavity or zone 2 with 100 mm clear cavity.
- Render to be correctly specified and applied to an appropriate backing material to minimise cracking, in accordance with BS 5262.

Third party certification
- Built-in cavity fill must have third party certification and be installed in accordance with manufacturers' instructions.
- Injected cavity fill must have third party certification and be installed under an approved surveillance scheme.
- External insulation must have third party certification for use on solid walls in that exposure zone.

Walls
Rain penetration through cavity and solid walls

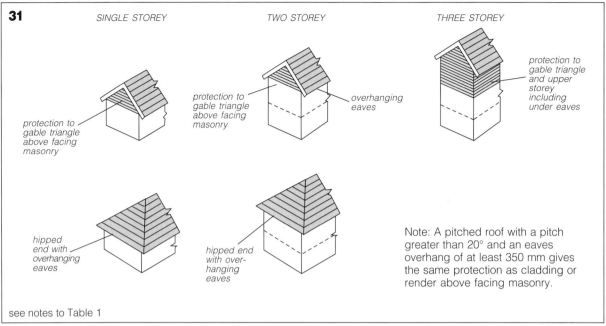

Interpretation of walls with impervious cladding or render above facing masonry

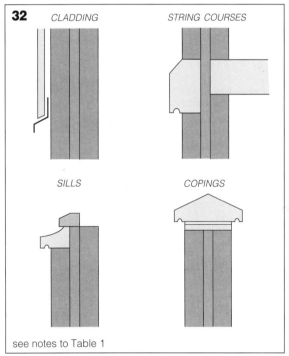

Masonry protected by overhangs

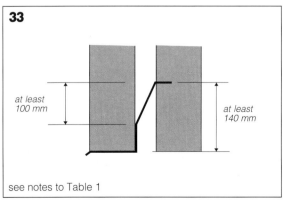

Minimum dimensions for cavity trays

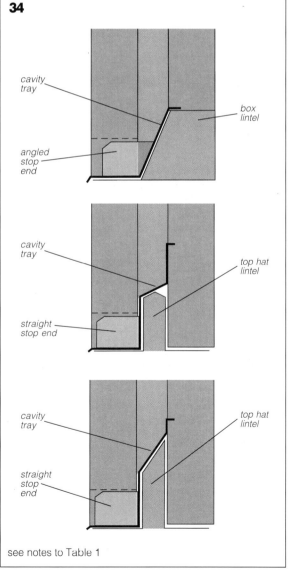

Cavity tray and stop end profiles above lintels

Walls

Masonry cavity walls

| Domestic | ✔ |
| Other sectors | ✔ |

CHARACTERISTICS OF THE CONSTRUCTION

This sub-section covers those technical risks common to all insulated walls which have a masonry outer leaf. The outer leaf may be of facing masonry or masonry with an additional protective finish such as render or tile hanging. The cavity may contain insulation.

R This guidance is appropriate when adding extensions or when upgrading the thermal performance of existing masonry walls.

ASSOCIATED TECHNICAL RISKS

- 3.2 Frost and/or sulphate attack on the outer leaf
- 3.3 Condensation at thermal bridges
- 3.4 Condensation within the construction
- 3.5 Increased heat loss due to air movement
- 3.6 Spread of fire gases in the wall cavity
- 3.7 Cracking of plaster due to blockwork shrinkage

Note: For guidance on avoiding rain penetration, refer to Risk 3.1 on page 21. Thermal bridging at the junctions of walls with other elements is covered in the Roofs, Windows and Floors sections as appropriate.

RISKS AND AVOIDING ACTIONS

3.2 Frost and/or sulphate attack on the outer leaf

The main factors affecting frost attack are the degree of exposure to wind driven rain combined with the frequency of freeze/thaw cycles and the frost resistance of the masonry.

The main factors affecting sulphate damage to mortar are the length of time the masonry is wet, the presence and type of soluble salts in the masonry units and the type of cement used for the mortar.

The changes in temperature and moisture content due to increased insulation are small and there is little evidence that insulated walls with a fairfaced outer leaf have a higher incidence of frost damage or sulphate attack than uninsulated walls. However, walls with a painted finish are more susceptible to frost and sulphate attack and the presence of full fill cavity insulation may increase the rate at which defective render deteriorates.

3.2 (a) For all walls, use masonry with sufficient frost resistance for the degree of exposure to wind driven rain and freezing temperatures. Clay bricks should be frost resistant (designation F in BS 3921) when used in exposure zone 4 on sites located over 90 m above sea level, receiving on average over 60 days of frost annually.

(b) Use a sulphate resisting cement in the base coat of the render and the brickwork mortar (see BS 5628: Part 3: Table 13) when considering rendering walls in exposure zones 4 and 3 onto clay bricks that have a soluble salt designation N (Normal as determined in BS 3921).

(c) Use only vapour permeable paint finishes on facing masonry and render, since impermeable finishes can prevent the wall drying out and increase the risk of frost damage and sulphate attack. The combination of a painted finish and full cavity fill can significantly increase the risk of frost damage.

3.3 Condensation at thermal bridges

If the continuity of wall insulation is broken by a dense element or an uninsulated component, the internal surface temperature may fall below dew point, causing mould growth and damage to wall decorations.

3.3 (a) Insulate behind recessed meter boxes with insulation with a thermal resistance of at least 0.6 m²K/W. Alternatively, avoid using recessed wall meter boxes.

(b) Insulate around or across masonry chimneys ensuring that no combustible insulation is closer than 200 mm to the flue (Figure **35**).

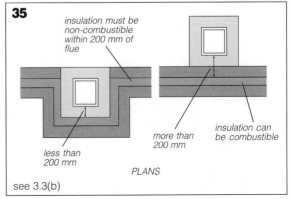

35

see 3.3(b)

Avoiding a thermal bridge at masonry chimneys

Thermal insulation: avoiding risks

Walls
Masonry cavity walls

3.4 Condensation within the construction

Condensation can cause damage if it forms on the inner surface of unventilated impervious cladding or if water vapour is restricted from passing through the construction by a vapour resistant layer on the cold side of the insulation.

3.4 (a) Ventilate behind any impervious cladding.

(b) Ensure aluminium foil on partial cavity insulation is perforated on the side facing the cavity to avoid condensation on the back of the foil.

3.5 Increased heat loss due to air movement

Heat loss is increased if cold air from the cavity is able to move behind partial fill insulation or through an air permeable inner leaf to the interior via holes for services or gaps around dry lining.

3.5 (a) Ensure that partial cavity insulation is held tightly against the inner leaf to avoid the movement of cold air behind the insulation.

(b) Seal the perimeter of the air space behind dry lining on external masonry walls by applying a continuous ribbon of plaster adhesive (not separate plaster dabs) at the ceiling, skirting, adjoining walls, window reveals and around holes made in the lining for services (Figure **36**).

3.6 Spread of fire gases in the wall cavity

Smoke and fire gases can spread into the roof from the wall cavity.

3.6 (a) Cavities filled with non-combustible insulation need no cavity barrier at the top. A cavity closer is useful to contain injected cavity wall insulation.

(b) Provide a cavity barrier at the top of unfilled cavities or cavities partially filled with insulation.

3.7 Cracking of plaster due to blockwork shrinkage

Concrete blocks shrink as they dry. If the blockwork is wet when plaster is applied, the subsequent shrinkage of the blocks can cause the plaster to crack.

3.7 (a) Minimise visible cracking by:
- waiting for blockwork to dry out fully before applying plaster and, if necessary, providing shrinkage control joints as recommended by the block manufacturer and in accordance with BS 5628: Part 3, OR
- using dry lining instead of plaster.

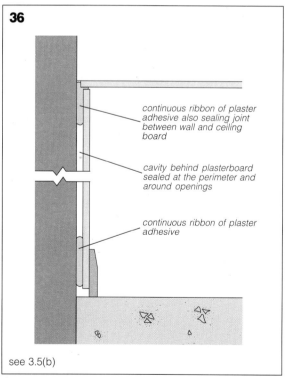

Sealing at the perimeter of dry lining

Walls

Masonry walls with internal insulation

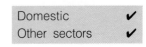
Domestic ✔
Other sectors ✔

CHARACTERISTICS OF THE CONSTRUCTION

Insulation is applied to the internal surface of masonry walls and finished usually with plasterboard.

Insulation which can be of mineral wool or plastics materials is usually bonded to the plasterboard finish as a composite board, but it can also be built up in layers with a vapour control layer between the insulation and the plasterboard lining. For new build, it may need to be used in conjunction with lightweight blockwork to achieve the required U-value. This may be the inner leaf of a cavity wall or a solid blockwork wall with cladding.

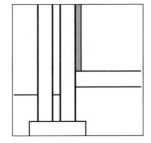

 This insulation method provides an energy efficient alternative to wet plaster when the internal surface of a complete external wall is to be renewed.

ASSOCIATED TECHNICAL RISKS

3.8 Increased heat loss due to air movement
3.9 Condensation within the construction
3.10 Summer condensation on a vapour control layer
3.11 Fire spread due to combustible insulation
3.12 Spread of fire gases in wall cavities
3.13 Condensation at thermal bridges
3.14 Risks associated with electrics

Note: For guidance on avoiding rain penetration, refer to Risk 3.1 on page 21. Thermal bridging at the junctions of walls with other elements is covered in the Roofs, Windows and Floors sections as appropriate.

RISKS AND AVOIDING ACTIONS

3.8 Increased heat loss due to air movement

Cold air moving through the masonry and then around or through holes in the insulated lining will cause an increase in heat loss.

3.8 (a) Seal the perimeter of air spaces behind insulated linings. In the case of insulating plasterboard, use continuous ribbons of plaster adhesive, as for dry lining (see 3.5(b)).

(b) Seal any holes made in the insulated lining for services.

3.9 Condensation within the construction

Condensation can cause damage within the construction if water vapour is unable to permeate through the wall or if moisture laden air reaches the cold surface of the masonry.

 3.9 (a) Provide a means of removing water vapour close to its source, eg install passive stack or mechanical extract ventilation in kitchens and bathrooms.

(b) Incorporate a vapour control layer on the warm side of the insulation by:

- using an insulating plasterboard with an integral vapour control layer, OR
- providing a separate vapour control layer, eg 500 gauge (0.12 mm) polyethylene, or a plasterboard lining with an integral vapour control layer when insulation is between timber battens. Timber in contact with the masonry should be preservative treated.

(c) Ensure that separate vapour control layers are continuous and lining boards with an integral vapour control layer are close butting, with joints sealed.

(d) Avoid running services on external walls whenever possible.

(e) Ensure neat cutting of composite boards or separate vapour control layers and plasterboard linings where services pass through and seal any gaps to prevent moist air reaching the cold masonry.

(f) As a general rule, condensation within the construction should be avoided if the vapour resistance of the layers on the warm side of the insulation is at least five times greater than that of the layers on the cold side. If this is not the case, for example due to the use of cladding with a high vapour resistance, assess the condensation risk using the calculation method in BS 5250.

Thermal insulation: avoiding risks   27

3.10 Summer condensation on a vapour control layer

In solid masonry walls with insulated linings orientated from ESE through South to WSW, sunshine falling on damp masonry can drive water vapour into the construction, causing condensation on the outside of the vapour control layer.

R 3.10 (a) For solid walls of facing masonry less than 250 mm thick, either:
- use a composite plasterboard laminate with an insulant of moderate or high vapour resistance, OR
- add 25 mm deep spacer blocks behind the timber battens when the internal insulation is built up in layers. Ventilate the space behind the insulation to the outside at top and bottom by means of openings through the masonry 750 mm² in area at 1.2 m centres. Ensure that the insulation is tight fitting between the timber battens to avoid increased heat loss. Seal the plasterboard lining at the top, bottom, all corners and at penetrations (Figure **37**).

(b) For new-build walls of solid masonry:
- use a rendered finish and adopt one of the internal insulation methods given in 3.10(a), OR
- protect the masonry with ventilated cladding (such as tile hanging or timber boarding) or with an outer masonry leaf with a clear cavity.

(c) Do not paint facing or rendered masonry with decorative finishes that are impermeable to water vapour.

3.11 Fire spread due to combustible insulation

Where combustible insulation is added to the internal face of a wall, it presents a fire hazard and should be protected by materials that can restrict its contribution to a fire.

3.11 (a) Protect combustible internal insulation with a facing material that can prevent fire involvement of the insulation, eg 12.5 mm plasterboard.

(b) Fix the facing material mechanically to the masonry to prevent premature collapse, as recommended by the insulation manufacturer.

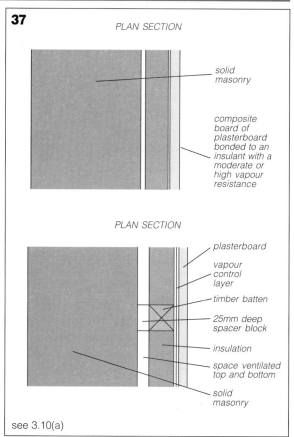

Internal insulation on solid walls

3.12 Spread of fire gases in the cavity behind insulated linings

Smoke and fire gases can spread to other elements through cavities behind insulated linings.

3.12 (a) Seal the junctions of insulating plasterboard laminates with floors, ceilings and adjoining walls using timber battens or continuous ribbons of plaster.

(b) Ensure that the ceiling plasterboard extends to the masonry wall and seal the joint with plaster adhesive (Figure **38**).

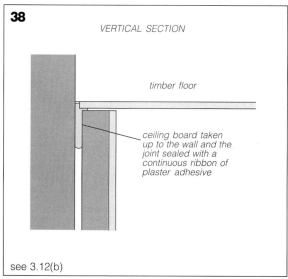

Sealing insulating plasterboard at the ceiling

Masonry walls with internal insulation

3.13 Condensation at thermal bridges

At junctions with separating walls and internal partitions, the continuity of internal insulation is broken, creating a potential thermal bridge.

3.13 (a) For solid external walls, or for the inner leaf of cavity walls, use masonry with the lowest possible density compatible with requirements for structural stability and flanking sound transmission.

SEPARATING WALL JUNCTIONS

(b) Avoid a thermal bridge at separating walls by:
- filling the external wall cavity with a strip of insulation so that it overlaps the internal insulation either side of the separating wall (this should be a flexible material such as mineral wool when required to restrict flanking sound transmission), OR
- choosing a separating wall specification which allows a medium or low density block to be used, OR
- lining the separating wall with plasterboard insulated with mineral wool, subject to sound insulation requirements (Figure **39**).

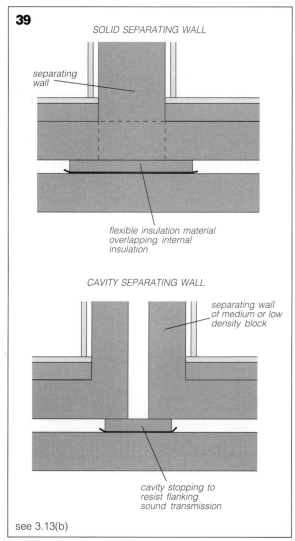

Junctions with separating walls

PARTITION JUNCTIONS

(c) Avoid a thermal bridge at partitions by:
- using insulating blockwork for partitions and for the external wall, OR
- taking the internal insulation across the external wall and constructing a timber stud partition up to it.

 (d) Use insulating blockwork for any new masonry partitions; tie to the external wall and finish with dry lining.

3.14 Risks associated with electrics

Electrical cables give off heat when in use. Where cables are covered by thermal insulation they may overheat, increasing the risk of short circuit or fire.

PVC sheathing to cables can have reduced life expectancy if in direct contact with expanded polystyrene insulants.

3.14 (a) Route cables where they will not be covered by insulation so that they can dissipate heat. If this is not possible, the cables may need to be increased in size, even when in conduit, especially where they serve cooker points and high output heaters (see Appendix A).

(b) Locate PVC insulated cables in conduit or away from direct contact with expanded polystyrene insulation.

(c) To avoid mechanical damage, specify metal conduit if the cable is within 50 mm of the plasterboard lining.

Walls

Masonry walls with external insulation

Domestic	✔
Other sectors	✔

CHARACTERISTICS OF THE CONSTRUCTION
Insulation is applied to the external surface of masonry walls and finished with cladding, render, proprietary surface coating, or tile hanging.

R This insulation method is appropriate for upgrading existing buildings:
- when the cavity is unsuitable for filling and internal insulation is not a viable option,
- when facing masonry is likely to deteriorate if left unprotected,
- when renewal of a rendered finish or cladding is being considered,
- when solid walls are to be insulated.

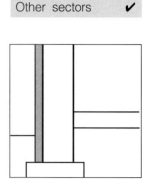

ASSOCIATED TECHNICAL RISKS
3.15 Impact damage and cracking of render
3.16 Fire propagation due to combustible insulation
3.17 Spread of fire in wall cavities
3.18 Condensation within the construction
3.19 Delamination of composite profiled sheet cladding panels

Note: For guidance on avoiding rain penetration, refer to Risk 3.1 on page 21. Thermal bridging at the junctions of walls with other elements is covered in the Roofs, Windows and Floors sections as appropriate.

RISKS AND AVOIDING ACTIONS

3.15 Impact damage and cracking of render

Rendering on insulation is subject to greater fluctuations in temperature and is more vulnerable to impact damage than when applied directly to masonry.

3.15 (a) Take precautions to minimise cracking by:
- reinforcing the render with a mesh,
- providing movement joints compatible with the system,
- using a light coloured finish,
- using a render incorporating a polymer and/or fibres.

3.16 Fire propagation due to combustible insulation

Where combustible insulation is added to the external face of a wall it presents a fire hazard unless protected.

3.16 (a) Protect combustible insulation boards with a render that can prevent active fire involvement of the insulant and fire propagation.

(b) Reinforce the render with mesh, mechanically fixed through the insulation to the masonry wall using non-combustible fixings, to prevent premature collapse of the render and incorporate cavity barriers across the insulation to comply with building regulations (Figure **40**).

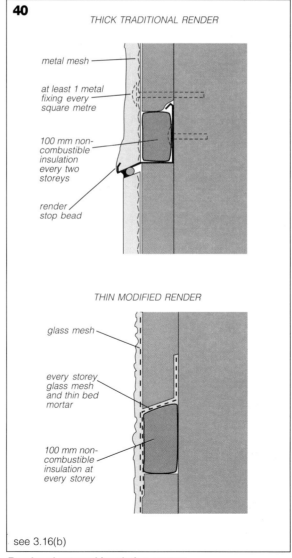

40

Rendered external insulation system

Masonry walls with external insulation **Walls**

3.17 Spread of fire in wall cavities

Smoke and fire gases can spread through cavities in the construction, particularly between insulation and cladding.

3.17 (a) Install a cavity barrier at the head of a cavity formed by overcladding (Figure **41**).

(b) Install cavity barriers between cladding and the structural wall at compartment walls and floors, and at each floor level of residential buildings as required by building regulations (Figure **41**).

(c) Ventilate the cavity behind impermeable cladding. This will require ventilation openings at the top and bottom of each void enclosed by cavity barriers (Figure **41**).

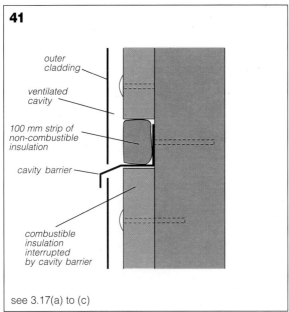

Cavity barriers with ventilated insulated cladding

3.18 Condensation within the construction

Condensation can occur on surfaces within the construction if water vapour from inside the building is unable to permeate through the external insulation system or be ventilated safely away.

3.18 (a) Use a breather membrane to BS 4016 behind tile hanging or similar cladding. Do not use polyethylene or building paper since they will restrict the passage of water vapour.

(b) Provide a drained and ventilated space behind PVC-U, metal sheet or other impermeable cladding.

3.19 Delamination of composite profiled sheet cladding panels

An inadequate bond during manufacture between the insulation and the outer sheet may be broken due to structural flexing or thermal movement of the outer sheet. This results in bowing and possibly condensation in the newly formed voids.

3.19 (a) Seek third party assurance from the manufacturer that there is an adequate bond between the insulation and the outer sheet of the cladding panel.

(b) Choose light colours where possible to minimise thermal movement.

(c) Design within recommended span limits for wind loading.

Thermal insulation: avoiding risks

Walls

Walls of timber framed construction

| Domestic | ✓ |
| Other sectors | ✓ |

CHARACTERISTICS OF THE CONSTRUCTION

The insulation is usually placed within the thickness of the timber framed structure between the outer sheathing and the inner lining. This insulation may be supplemented by an additional lining of thermal insulation or insulated sheathing to reduce the heat loss through the timber studs.

 This construction method is appropriate for extensions.

ASSOCIATED TECHNICAL RISKS

3.20 Condensation within the construction
3.21 Risks associated with electrics

Note: For guidance on avoiding rain penetration, refer to Risk 3.1 on page 21. Thermal bridging at the junctions of walls with other elements is covered in the Roofs, Windows and Floors sections as appropriate.

RISKS AND AVOIDING ACTIONS

3.20 Condensation within the construction

If water vapour from inside the building is allowed to pass through the insulation to the cold part of the structure, it may condense. If this occurs for extended periods, the timber structure may be at risk of decay.

3.20(a) Incorporate a vapour control layer on the inside face of the timber studs by:
- providing a separate vapour control layer, eg 500 gauge (0.12 mm) polyethylene behind the plasterboard lining, OR
- using a plasterboard lining with an integral vapour control layer (Figure **42**).

(b) Ensure that the separate vapour control layer is continuous or that lining boards with an integral vapour control layer are close butting over framing supports (Figure **42**).

(c) Avoid running services in external walls whenever possible.

(d) When it is not possible to avoid services in the external wall:
- ensure neat cutting of the separate vapour control layer, or plasterboard linings with an integral vapour control layer, around services and seal any residual gaps, OR
- route service runs within the insulation of a composite plasterboard laminate on the warm side of the vapour control layer (Figure **42**).

(e) When insulation is introduced on the warm side of the vapour control layer, ensure that the thermal resistance of the construction on the cold side of the vapour control layer is at least three times greater than the resistance of the construction on the warm side.

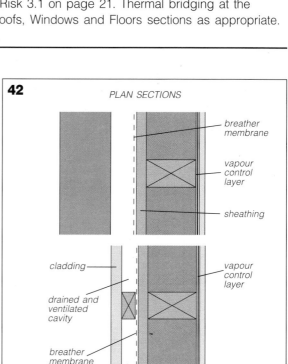

see 3.20(a), (b), (c), (d) and (g)

Typical timber frame construction methods

(f) Ensure that the space behind cladding is drained and ventilated (Figure **42**).

(g) As a general rule, condensation within the construction should be avoided if the combined vapour resistance of the layers on the warm side of the insulation is at least five times greater than that of the layers on the cold side. If this is not the case, for example due to the use of cladding or sheathing with a high vapour resistance, assess the condensation risk using the calculation method in BS 5250.

Thermal insulation: avoiding risks

Walls of timber framed construction

3.21 Risks associated with electrics

> *Electrical cables give off heat when in use. Where cables are covered by thermal insulation they may overheat, increasing the risk of short circuit or fire.*
>
> *PVC sheathing to cables can have reduced life expectancy if in direct contact with expanded polystyrene insulants.*
>
> *Unprotected cables can be punctured by nails if close to the wall surface.*

3.21(a) Route cables where they will not be covered by insulation, so that they can dissipate heat. If this is not possible, the cables may need to be increased in size, even when in conduit. The circuits most likely to be affected are radial circuits serving cookers, immersion heaters, shower units, socket outlets and high wattage heaters.
(see Appendix A).

(b) Locate PVC insulated cables in conduit or away from direct contact with expanded polystyrene insulation.

(c) To avoid mechanical damage, specify metal conduit if the cable is within 50 mm of the plasterboard lining.

Walls

Quality control checks for walls

MASONRY CAVITY WALLS

- Is the wall construction, including the cavity width and the insulation, suitable for the wall exposure?
- Are the masonry units, mortar and any finishes as specified, particularly in respect of frost and sulphate resistance?
- Is the cavity width within specified limits?
- Are all mortar joints, particularly perpends, filled and are all weepholes clear?
- Is the outer leaf built ahead of the inner leaf when using full fill batts?
- Is the inner leaf built ahead of the outer leaf when using partial cavity insulation?
- Is the cavity free of mortar droppings and other debris?
- Are all dpcs and cavity trays as specified?
- For built-in cavity fill:
  – have third party certificate and manufacturers' installation procedures been followed?
  – is insulation tightly butted and are off-cuts correctly aligned?
  – have mortar droppings been removed from exposed edges of insulation?
  – is partial fill insulation held tightly against inner leaf?
- For injected cavity fill:
  – has the third party approved installer accepted the wall as suitable for filling?
  – is installation in accordance with relevant standards or third party certification and associated surveillance schemes?
- Has insulation been included to reduce thermal bridging at meter boxes and chimneys?

MASONRY WALLS WITH INTERNAL INSULATION

- Are the masonry type and thickness suitable for the wall exposure?
- Are the masonry units, mortar and any finishes as specified, particularly in respect of frost and sulphate resistance?
- Do continuous ribbons of plaster seal the space behind insulated plasterboard linings at the perimeter of walls and openings, and around holes in plasterboard?
- Is a continuous vapour control layer with lapped and taped joints fitted behind plasterboard linings, or is it incorporated as a sandwich in composite boards?

- Are external paint finishes permeable and is ventilation provided behind any external cladding?
- Is ventilation provided behind masonry where specified?
- Has insulation been included to reduce thermal bridging at wall and partition junctions?
- Are adhesive fixed composite boards stabilised by mechanical fixings?
- Is contact avoided between PVC insulated cables and polystyrene and are electric cables rated correctly where in contact with any insulation?

MASONRY WALLS WITH EXTERNAL INSULATION

- Is the cladding suitable for the wall exposure?
- Has the external insulation system been applied in accordance with third party certification by approved installers?
- Are fire barriers installed correctly at the recommended centres?
- Is there effective ventilation and drainage behind ventilated overcladding, particularly at fire barriers?
- Is the membrane behind tile hanging or similar cladding of a vapour permeable type?

WALLS OF TIMBER FRAMED CONSTRUCTION

- Are the cladding and the clear cavity width suitable for the wall exposure?
- Is the vapour control layer continuous and neatly cut and sealed where services penetrate it?
- Does insulation fully fill the spaces between the timbers?
- Are cables rated correctly when in contact with insulation and are they protected from piercing by fixings?
- Is the cavity free of mortar droppings and other debris?
- Are all dpcs and cavity trays as specified?
- Is the space behind cladding drained and ventilated?

4 Windows

Window to wall junctions	35
Double glazed windows	39
Quality control checks for windows	42

Window to wall junctions

CHARACTERISTICS OF THE CONSTRUCTION

This sub-section covers the technical risks which can occur where window or external door frames meet the surrounding masonry wall.

 This guidance is appropriate when replacement windows are installed in an existing structural opening or when the internal plaster finish of the external wall is to be replaced with internal insulation.

Domestic ✓
Other sectors ✓

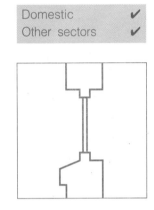

ASSOCIATED TECHNICAL RISKS

4.1 Condensation and mould at thermal bridges
4.2 Increased heat loss due to air movement
4.3 Mould on reveals due to contact with metal window frames

RISKS AND AVOIDING ACTIONS

4.1 Condensation and mould at thermal bridges

If a dense part of the construction interrupts the continuity of the insulating layer, the internal surface temperature may fall locally below dewpoint, causing mould growth and damage to wall decorations.

4.1 (a) When a window or door frame is set back behind the inner face of a dense outer masonry leaf, it should overlap an insulated closer or insulating block closer by a minimum dimension which relates to the thermal characteristics of the closer (Figures **43** and **44**).

(b) Set the frame back to overlap the insulation by at least 25 mm when it is in the form of:

- a proprietary closer which incorporates insulation material with a thermal conductivity no greater than 0.04 W/mK, OR
- a strip of insulation with a thermal resistance of at least 0.25 m^2K/W between the outer leaf and the blockwork closer or inner leaf (Figure **43**).

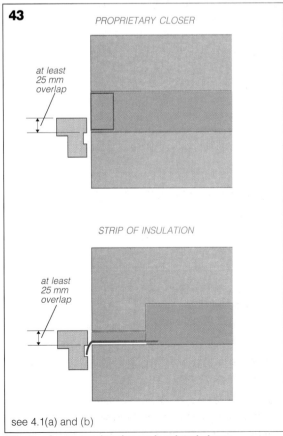

see 4.1(a) and (b)

Window frame overlapping an insulated closer

Windows

Window to wall junctions

4.1 (c) Set the frame back to overlap a blockwork cavity closer:
- by at least 45 mm if the thermal conductivity of the block is 0.2 W/mK or less, OR
- by the full frame width, eg in a rebated or checked reveal, if the thermal conductivity of the block is 0.3 W/mK or less (Figure **44**).

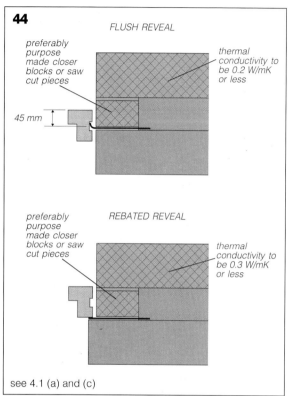

see 4.1 (a) and (c)

Window frame overlapping a blockwork closer

(d) Apply insulation to the cavity closer at the jamb and beneath the window board at the sill using insulating material with a thermal resistance of at least 0.25 m²K/W where:
- the window frame is set forward within the outer masonry leaf (irrespective of the thermal characteristics of the material used for the cavity closer),
- the wall is insulated with internal insulation (irrespective of the position of the frame in the reveal) (Figure **45**).

(e) Add an insulated lining to the soffit of lintels when:
- the window frame is set forward within the outer masonry leaf,
- the wall is insulated with internal insulation,
- they are made of steel and have a continuous lower web (Figure **46**).

(f) Consider whether the position of trickle ventilators in the frame is affected by the presence of an insulated soffit lining.

 (g) Follow the guidance given in 4.1(a) to (f) when replacing windows or substituting insulating plasterboard for plaster as an internal wall finish.

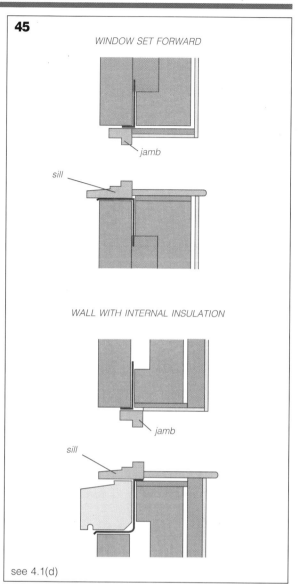

see 4.1(d)

Insulation to reveals at jambs and sills

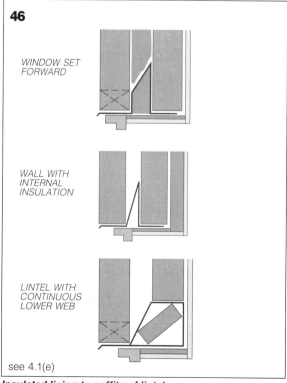

see 4.1(e)

Insulated lining to soffits of lintels

Window to wall junctions **Windows**

4.1 (h) Fill as much as possible of the space between the frame and the wall with insulation, such as expanding foam or mineral wool (Figure **48**).

(i) To avoid condensation risk at the corners of a reveal, ensure that insulated closers between inner and outer masonry leaves at jambs and sills are either in the same plane (Figure **47**), or overlap against one another.

4.2 (b) Allow sufficient tolerance in building openings to allow for window installation, making the seals and filling the space between the frame and the wall with insulation (Figure **48**).

It is preferable to build the opening in the inner masonry leaf wider than that in the outer leaf to create sufficient space to insert insulation into the joint and make an effective secondary seal.

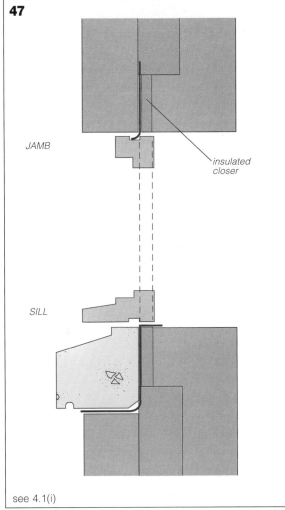

see 4.1(i)

Insulation in the same plane at closers

4.2 Increased heat loss due to air movement

Failure to make an airtight seal between the window frame and the walling can result in unacceptable draughts and unnecessary heat loss.

4.2 (a) Seal the joint between the frame and the wall with an outer primary seal and an inner secondary seal using:
- sealant pointing on the outside and expanding foam or plaster on the inside, OR
- impregnated foam strip on the outside and sealant pointing on the inside, OR
- sealant bedding on the outside and plaster adhesive on the inside (Figure **48**).

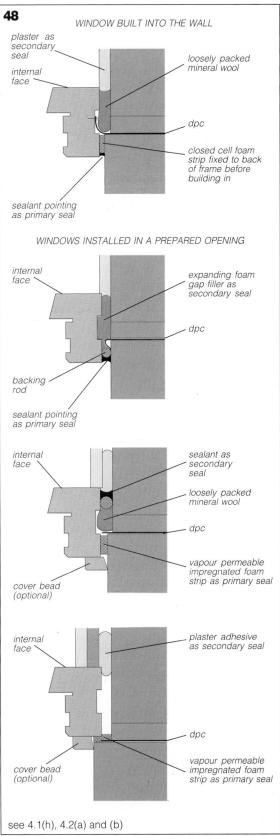

see 4.1(h), 4.2(a) and (b)

Examples of sealing at window to wall junctions

Windows

4.3 Mould on reveals due to contact with metal window frames

Metal windows can suffer badly from condensation and, if they abut absorbent finishes such as plaster, their high thermal conductivity can be the cause of mould growth at window reveals.

4.3 (a) For preference, select metal windows which incorporate thermal breaks (Figure **49**).

(b) For metal windows without thermal breaks:
- install the window in a durable timber sub-frame and incorporate a channel to collect temporarily any condensation run-off from the metal frame, OR
- fill the space between the metal frame and the wall with expanding polyurethane foam (Figure **49**).

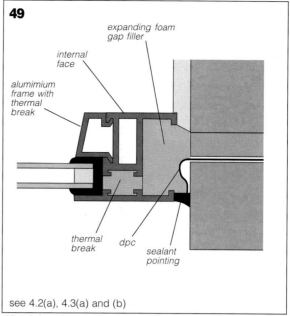

see 4.2(a), 4.3(a) and (b)

Aluminium frame with thermal break and foam sealing to wall

Windows

Double glazed windows

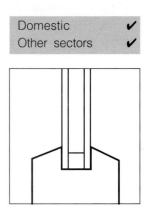

| Domestic | ✔ |
| Other sectors | ✔ |

CHARACTERISTICS OF THE CONSTRUCTION

Under revised building regulations, double glazing will become the benchmark for windows, rooflights and some glazed external doors. Double glazed windows may use sealed units or be formed by double windows.

R Double glazing may be provided by replacement windows or by adding secondary glazing to a single glazed window opening.

ASSOCIATED TECHNICAL RISKS

4.4 Failure of edge seals in double glazing units
4.5 Deformation of window frames
4.6 Condensation on the inside surface of the external pane of double windows
4.7 Inability to escape in case of fire
4.8 Condensation due to restricted ventilation

RISKS AND AVOIDING ACTIONS

4.4 Failure of edge seals in double glazing units

Edge seals can be defective if there is inadequate quality control in manufacture. They may also deteriorate if they are exposed to sunlight or if water is trapped against the edge seal for prolonged periods. This deterioration results in condensation within the double glazing unit. There may also be incompatiblity between the glazing compound and the edge seal material or decorative finish, leading to failure of the edge seal.

4.4 (a) For best results, choose a window that is factory glazed.

DOUBLE GLAZING UNITS
  (b) Use sealed units tested to BS 5713 and ensure that they are Kitemarked. Use dual seal units (Figure **50**).

  (c) Check with the manufacturer of the sealed units that the material for the edge seal is compatible with the proposed glazing compounds.

  (d) Protect the edge seal from sunlight by ensuring that the rebate is of sufficient depth, including spacers, for the edge seal of the double glazing unit to be below the sight line of the rebate upstand and the glazing bead (Figure **51, 53** and **54**).

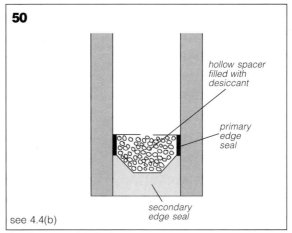

Glazing unit with dual seal

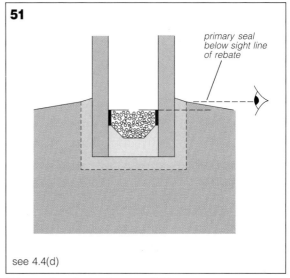

Edge seal protected from sunlight

Thermal insulation: avoiding risks

Windows
Double glazed windows

GLAZING METHODS

4.4 (e) Avoid methods that rely on a complex installation procedure or frequent maintenance to prevent premature failure of the edge seal. *Glazing techniques for insulating glass units* by the GGF and BS 8000:Part 7 give more detailed advice on preferred glazing methods.

(f) Use a drained and/or ventilated glazing method whenever possible for PVC-U, aluminium and timber windows (Figures **52** and **53**).

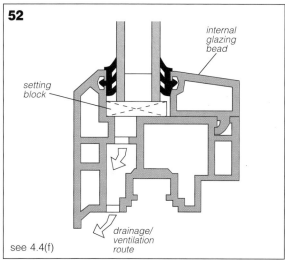

PVC-U frame with drained and ventilated glazing method

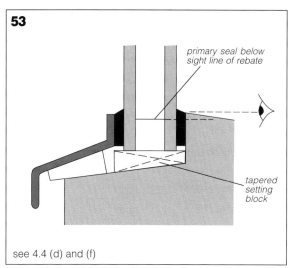

Timber frame with drained and ventilated glazing method

(g) Use a fully bedded method for timber and steel windows when a drained and/or ventilated method is not suitable. Use a sealant alone or preformed sealant strips with sealant capping. Specify that there should be no voids within or between the glazing materials (Figure **54**).

(h) Ensure that water is shed away from the glass by:
- tooling the sealant or sealant capping to a smooth chamfer, OR
- trimming strip sealants to give a sloping top edge (Figure **54**).

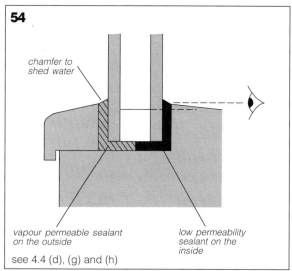

Timber frame with fully bedded sealing method

4.4 (i) Ensure that the glazing compounds are compatible with the decorative finish for the window. **Do not use putty.**

4.5 Deformation of window frames

Some lightweight window frames are not robust enough for the extra weight of double glazing units. Frames may deflect repeatedly under wind gusting, causing glazing units to crack or seals to be broken. Frames especially of opening lights may also distort if the glazing units are not properly supported or restrained.

4.5 (a) Use windows whose frames, hinges and opening mechanisms are designed to carry the weight of the specified double glazing units. Do not exceed manufacturers' recommended limits on sash weights and dimensions.

(b) Provide support and restraint for the double glazing units within the frame by using:
- setting blocks at least 3 mm thick on the horizontal glazing platform,
- location blocks, in vertical and top rebates to prevent movement of the unit in the frame when windows are open or closed, positioned according to the hinge arrangement,
- distance pieces at least 3 mm thick where necessary to prevent displacement of the glazing compound due to wind pressure on the glass. Distance pieces should be opposite each other on both sides of the unit. They are not necessary when using load bearing strip compounds.

(c) Ensure that the frame rebate is of sufficient size to allow for:
- clearance around the unit to accommodate setting and location blocks,
- tolerance in the size of the double glazing unit.

Double glazed windows *Windows*

4.6 Condensation on the inside surface of the external pane of double windows

If water vapour produced within the building condenses on the inside surface of the external pane of double windows, it can cause unacceptable fogging and eventually deterioration of lower frame members.

4.6 (a) Seal the inner window and seal any gaps around the frame to restrict the movement of moist internal air into the cavity.

(b) Ventilate the space between the double windows to the outside. The gap around the undraughtstripped opening lights of the outer window usually provides sufficient ventilation, without reducing significantly the thermal insulation performance.

R **(c)** Seal secondary glazing and ventilate the space between the panes to the outside, for example, by leaving top and bottom edges of opening lights in the existing windows undraughtstripped.

4.7 Inability to escape in case of fire

If double glazed windows are not easily opened in the event of a fire, lives may be at risk.

4.7 (a) Select new double glazed windows which have opening lights with easily operated opening mechanisms of a size and design to allow escape in an emergency.

R **(b)** Ensure that secondary glazing can be opened or removed quickly to allow escape in an emergency.

4.8 Condensation due to restricted ventilation

Secondary glazing to existing windows will restrict ventilation, leading to condensation and mould growth.

R **4.8 (a)** Provide a means of achieving natural ventilation to a room when installing secondary glazing by:
- ensuring that existing trickle ventilators remain effective, OR
- extending the existing trickle ventilator path to the inside, OR
- installing controllable background ventilation elsewhere in the external wall.

Thermal insulation: avoiding risks

Quality control checks for windows

WINDOW TO WALL JUNCTIONS

- Does the window frame overlap insulating cavity closers and by the required amount to avoid condensation risk?
- Has an insulated lining been used to the reveal when the window frame is set forward solely within the outer leaf?
- Has an insulated soffit lining been used beneath steel lintels with a continuous lower web?
- Have trickle vents been located so as not to be affected by an insulated soffit lining?
- Are window/wall junctions insulated and is the joint sealed both externally and internally?
- Have metal windows without thermal breaks been installed in timber sub-frames?

DOUBLE GLAZED WINDOWS

- Are sealed units Kitemarked BS 5713 on the spacer bar?
- Is the rebate upstand deep enought to protect the edge seal from sunlight?
- Are setting blocks, location blocks and distance pieces correctly installed?
- Are the internal windows of double windows sealed and is the space between the panes ventilated to the outside?
- Do double glazed windows allow effective escape in fire?
- Do the windows allow effective ventilation by natural means, or is it provided by other methods?

5 Floors

Concrete ground floors insulated below the structure	43
Concrete ground floors insulated above the structure	47
Concrete ground floors insulated at the edge	52
Suspended timber ground floors	55
Concrete and timber upper floors	58
Quality control checks for floors	61

Concrete ground floors insulated below the structure

CHARACTERISTICS OF THE CONSTRUCTION

Insulation is usually laid horizontally below the whole area of the floor.

Insulation should be rigid boards with a high moisture resistance and compressive strength and, if below the damp proof membrane (dpm), resistant to any contaminants in the ground or fill.

Concrete floors insulated in this way are cast in-situ and can be:
- supported on the ground or fill (ie not built into the surrounding walls), or
- supported by the surrounding walls as reinforced suspended slabs (ie built into the walling).

The dpm may be above or below the concrete; if below the concrete, the dpm can be above or below the insulation. On landfill sites and on ground containing high levels of sulphates, an additional membrane beneath the concrete may be needed to separate the concrete from ground contaminants when the damp proof membrane is to be above the concrete.

 These recommendations apply when it is proposed to replace an existing concrete floor or a timber suspended floor with one of in-situ concrete.

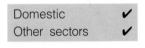

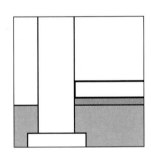

ASSOCIATED TECHNICAL RISKS

5.1 Reduced structural stability and thermal performance
5.2 Damage during construction
5.3 Condensation at thermal bridges
5.4 Damage from moisture in the floor
5.5 Risks associated with services

Note: Where necessary, dpms and tanking have been omitted in the figures to the Floors section, since there are several options for their location.

Thermal insulation: avoiding risks

Floors
Concrete ground floors insulated below the structure

RISKS AND AVOIDING ACTIONS

5.1 Reduced structural stability and thermal performance

As ground supported slabs are cast directly onto the insulation, deformation of the insulation could lead to initial subsidence, unexpected stresses in the floor slab and a reduction in thermal performance. Deformation may be caused by compression or degradation where insulants react chemically with contaminants in the ground or fill.

5.1 (a) To avoid degradation of the insulation:
- use a type with low water absorption and high resistance to any expected chemicals in the ground or fill, OR
- place the insulation on a dpm or separating membrane (Figure **55**).

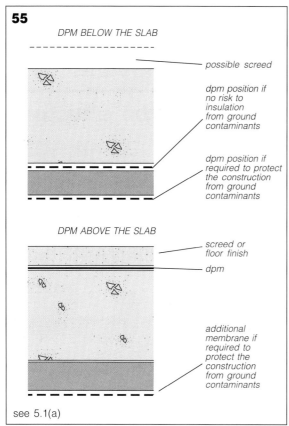

Alternative dpm positions

GROUND SUPPORTED SLABS

(b) Use an insulant with sufficient compressive strength to resist the weight of the slab, the anticipated floor loading as well as any possible overloading due to stored materials. Ensure that hardcore is well compacted and covered with sand blinding to give an even support for the insulation. Provide separate foundations for internal loadbearing walls.

REINFORCED SUSPENDED SLABS

(c) Ensure that the insulation has sufficient compressive strength to act as permanent shuttering for the concrete floor slab.

5.2 Damage during construction

Unless care is taken when laying the floor, the dpm and the insulation materials can become displaced or damaged.

5.2 (a) Lay the insulation boards and dpm (if below the slab) immediately before placing the concrete.

(b) Consider laying the dpm above the slab if appropriate for the floor finish.

(c) Ensure that the insulation boards fit tightly together and are weighed down as necessary before and after laying the dpm.

(d) Take extra care before and during laying concrete to prevent damaging the dpm with sharp objects.

GROUND SUPPORTED SLABS

(e) When tamping, protect the insulation and the dpm where they turn up vertically to meet the wall dpc, for example, with a temporary timber board (Figure **56**).

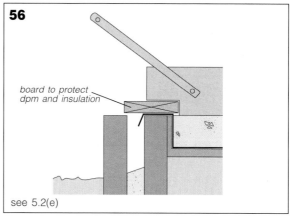

Protection to dpm and insulation when tamping

(f) To avoid damaging the dpm and insulation when power trowelling:
- protect the edge of the dpm and insulation with a board and hand trowel close to the edge of the floor (Figure **57**), OR
- lay the dpm above the slab (if compatible with the floor finish) later in the construction process.

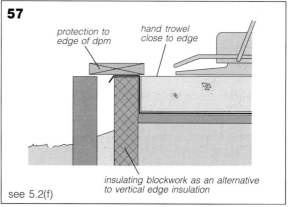

Precautions when power trowelling

Concrete ground floors insulated below the structure — **Floors**

REINFORCED SUSPENDED SLABS

5.2 (g) Ensure a firm base for the dpm and insulation to avoid the membrane being torn where it meets the external wall.

(h) Take care not to puncture the dpm when placing the reinforcement.

5.3 Condensation at thermal bridges

Where the continuity of the insulation is broken at junctions between the floor and external walls or load bearing internal walls, a thermal bridge occurs and there is the risk of surface condensation.

5.3 (a) Start cavity insulation below damp proof course (dpc) level, and at least to the same level as the base of the floor slab (Figures **58** and **59**).

(b) Use insulating blockwork for the inner leaf of a cavity wall below dpc, if structurally acceptable and suitable for the ground conditions (Figure **58**).

GROUND SUPPORTED SLABS

(c) Use a vertical strip of insulation at the perimeter of the slab:
- to provide an overlap with cavity insulation where the inner leaf below dpc is not insulating blockwork (Figure **59**), OR
- to link with internal wall insulation (Figure **60**).

(d) Use a strip of insulation at the perimeter of the screed to maintain continuity between the slab insulation and internal wall insulation (Figure **61**). Ensure that the top of the perimeter insulation is protected by the skirting.

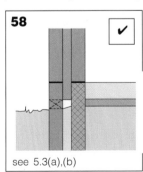
see 5.3(a),(b)

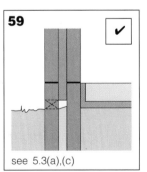
see 5.3(a),(c)

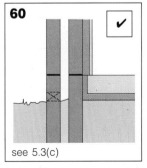
see 5.3(c)

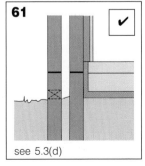
see 5.3(d)

Note: Where necessary, dpms and tanking have been omitted in the figures to the Floors section, since there are several options for their location.

5.3 (e) For internal loadbearing walls:
- place a vertical strip of insulation against internal loadbearing walls for a distance of 1 metre from any junction with an external wall (Figure **62**), OR
- use insulating blockwork below dpc (Figure **63**).

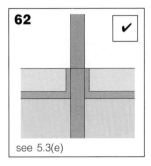
see 5.3(e)

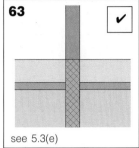
see 5.3(e)

(f) For stepped separating walls:
- choose a sound insulation specification which allows the use of insulating blockwork (subject also to structural requirements) (Figure **64**) OR,
- insulate within the cavity from the higher of the two floors down to the foundation with resilient insulation (eg mineral wool).

REINFORCED SUSPENDED SLABS

(g) Avoid using suspended slabs in conjunction with internal insulation, since it is difficult to avoid a thermal bridge (Figure **65**).

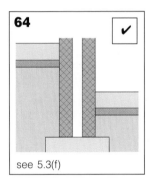
see 5.3(f)

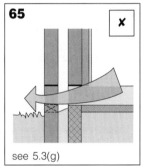
see 5.3(g)

(h) Use suspended slabs in conjunction with cavity insulation only if it is possible to support the floor on insulating blockwork. Otherwise, insulate above the floor. Check that insulating blockwork is suitable for the ground conditions and acceptable structurally (Figure **66**).

(i) At stepped separating walls, insulate within the cavity down to foundation level, as described in 5.3(f), (Figure **67**).

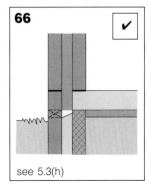
see 5.3(h)

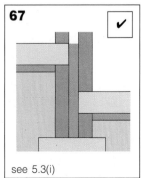
see 5.3(i)

Floors
Concrete ground floors insulated below the structure

5.4 Damage from moisture in the floor

Moisture sensitive floor finishes can be damaged by residual moisture in the concrete.

5.4 (a) Locate the dpm above the slab if there is any doubt whether the slab will have dried out to an acceptable moisture content for the chosen floor finish. A 100 mm in-situ concrete slab can take up to 6 months to dry sufficiently to receive moisture sensitive flooring. Wetting by rainwater can increase this time greatly.

(b) Use a sheet, or a hot or cold applied dpm above the slab and ensure it links with the wall dpc.

(c) If there is no doubt that the concrete will dry out, lay the dpm below the slab.

GROUND SUPPORTED SLABS

(d) Where the dpm is to be below the slab, lay it either **A** above the insulation if the insulation is resistant to ground contaminants, or **B** below the insulation if the insulation is not resistant to ground contaminants or if the whole of the floor needs to be protected, eg on landfill sites (Figure **68**).

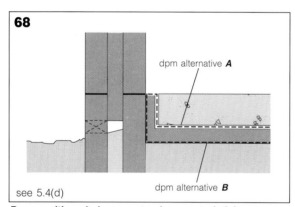

Dpm positions below a ground supported slab

REINFORCED SUSPENDED SLABS

(e) Locate the dpm above a reinforced suspended slab and link with the wall dpc (Figure **69**).

(f) Ensure that the insulation material is resistant to water and contaminants (Figure **69**).

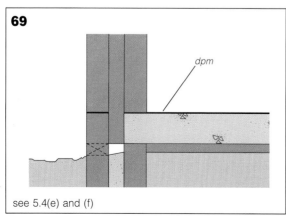

Dpm located above a reinforced suspended slab

5.5 Risks associated with services

Water supply pipes below floor insulation can be damaged by frost. Services within the floor can be damaged by fixings or frost. They can also cause dampness when they puncture damp proof membranes. Some water services need to be accessible for replacement or repair.

5.5 (a) Avoid laying services in the floor slab if the dpm is above the slab, as it is not possible to maintain the integrity of the membrane.

(b) Insulate the cold water supply pipe throughout its length when it enters the building less than 750 mm from the outside face of the external wall (Figure **70**). Using an insulant with a thermal conductivity of 0.035 W/mK (eg foamed plastics), a minimum wall thickness of 25 mm is needed for 15 mm (outside diameter) pipes and a 19 mm wall thickness for 22 to 28 mm pipes. For other types of insulation and pipe sizes, refer to Appendix B.

(c) Seal the damp proof membrane where pipes pass through the floor (Figure **70**).

(d) Make a rodent proof, gas tight seal where the water supply pipe enters the building using expanding foam with cement mortar pointing (Figure **70**).

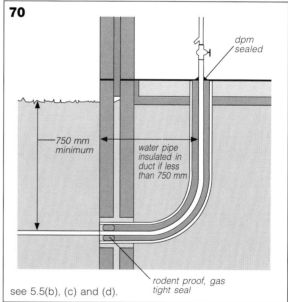

Insulation to a cold water supply through floors in contact with the ground

(e) Run water supply pipes in a duct with an accessible cover, either in a power floated slab or in a screed.

(f) Insulate central heating pipes and provide at least 25 mm cover of screed or, if in a power floated slab, run the pipes in a duct.

(g) Provide at least 25 mm cover to gas pipes and electrical conduit, either in a power floated slab or a screed. Use metal electrical conduit if the cover is less than 25 mm.

Concrete ground floors insulated above the structure

CHARACTERISTICS OF THE CONSTRUCTION

Insulation is used above the structure with precast concrete beam and block suspended floors, and with in-situ floors when the dpm is above the slab.

The insulation material should be rigid and suitable for the loading requirements. The insulation can be below a timber based flooring system or below a screed.

Insulated timber based flooring systems can be laid as:
- composite panels in which the insulant is bonded to the timber based board,
- loose laid systems in which the insulant and the boards are installed separately,
- timber battened systems in which the battens support the flooring panels and insulation is laid between the battens.

Domestic ✔
Other sectors ✔

This insulation method is relevant if a timber suspended floor is to be replaced by a precast concrete system or an in-situ concrete slab, or if it is convenient to lay insulation above an existing concrete floor.

ASSOCIATED TECHNICAL RISKS

5.6 Failure in service
5.7 Damage to the insulation during construction
5.8 Damage from moisture
5.9 Condensation at thermal bridges
5.10 Risks associated with services

RISKS AND AVOIDING ACTIONS

5.6 Failure in service

Failure in service can result from uneven support, incorrect installation of flooring panels or insufficient thickness and strength of screed.

5.6 (a) Ensure that the surface of the concrete slab or precast flooring is flat (up to 5 mm under a 3 metre straight edge is acceptable), clean and free from mortar or plaster droppings.

A roughly tamped concrete surface is not suitable without being levelled, whereas a smooth floated surface will be suitable without special treatment. Some precast plank or beam and block floors will provide a suitable base if the levelling layer is floated off evenly. Others are less accurate, for example, where beams with a camber meet at right angles and require a levelling screed.

(b) Choose an insulation board with adequate compressive strength for the intended loading when it supports timber based panels or screed.

TIMBER BASED FLOORING

(c) Fit tongued and grooved panels tightly together and glue all joints evenly with a PVA adhesive. Wedge the panels temporarily at the perimeter until the glue has set.

Failure to achieve an even surface to support the insulation can result in the glued joints breaking apart, causing a squeaky floor.

(d) Provide an expansion gap of at least 10 mm between timber-based particle board flooring panels and perimeter walls or other rigid upstands. When the width of floor between walls or upstands is more than 10 metres, allow a total expansion gap of 2 mm per metre width, distributed between opposite edges of the floor. Some manufacturers recommend an extra gap at an intermediate position across the floor.

Floors — Concrete floors insulated above the structure

5.6 (e) Support edges of the panels at door openings on preservative treated timber battens. Ensure that the battens are on a firm and level base and fix a strip of flooring to the battens as a threshold. Allow a gap each side of the threshold for movement in the flooring panels.

Use additional support battens where extra floor loading is anticipated and the exact position is known, eg beneath kitchen fittings, equipment and sanitary fittings (Figure **71**).

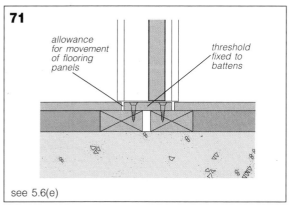

see 5.6(e)

Extra support and allowance for movement at doorways

(f) When required, use a levelling screed to ensure that the battens of a timber battened system are true and level. Do not attempt to fix the flooring to the battens through resilient insulation material, as this will create an uneven floor surface.

(g) Lightly sand and sweep chipboard floors to make the surface suitable for thin flooring such as flexible PVC sheet. Do not wash or scrub with water.

SCREED FINISH

(h) Use a cement:sand screed of 1:3 to 1:4.5 mix by weight, laid to a minimum thickness of 65 mm for domestic construction and 75 mm elsewhere.

Ensure that the screed is thoroughly compacted throughout its thickness, particularly above resilient insulation. If necessary, place and compact in two layers.

(i) Lay a separating layer of building paper (grade BIF to BS 1521) or 500 gauge (0.12 mm) polyethylene sheet above fibrous insulation to prevent the wet screed soaking into the insulation (Figure **72**).

(j) Tape the joints of rigid insulation boards so that wet screed does not penetrate the joints (Figure **72**).

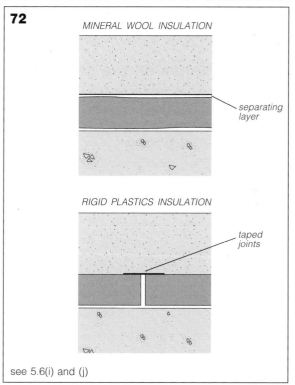

see 5.6(i) and (j)

Construction of insulated screeds

5.7 Damage to the insulation during construction

Insulation may be damaged by chemical reaction with solvents in some liquid damp proof materials and adhesives. When screeds are being laid, the insulation and the separating layer may be damaged.

5.7 (a) Check the chemical compatibility of liquid dpm materials with the insulation material. Some pitch and bitumen based materials can degrade polystyrene.

Ensure that the insulation is not laid until the solvents from liquid dpm materials have fully evaporated.

(b) Use boards to protect the insulation when barrowing screed material and tipping the wet mix into place (Figure **73**).

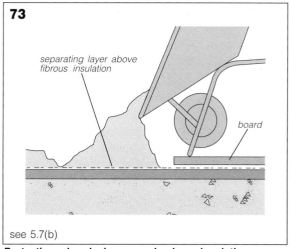

see 5.7(b)

Protection when laying screeds above insulation

Concrete floors insulated above the structure **Floors**

5.8 Damage from moisture

Moisture sensitive floorings and timber based flooring panels, particularly those of chipboard, can be damaged by water vapour emanating from moisture in the floor construction.

5.8 (a) For suspended floors of precast construction, introduce a separate 500 gauge (0.12 mm) polyethylene vapour control layer above the floor to protect timber based flooring panels and, when necessary, moisture sensitive floor finishes (Figures **75** to **77**).

Although precast concrete floors are essentially dry, protection is needed against residual moisture from construction or exposure to the weather.

(b) For in-situ concrete floors:
- lay the dpm above the slab, OR
- locate the dpm below the floor and introduce a 500 gauge (0.12 mm) polyethylene vapour control layer above the slab to protect timber based flooring panels and, when necessary, moisture sensitive floor finishes.

(c) To prevent residual dampness in the wall affecting the edge of the flooring:
- apply the liquid dpm up the wall (Figure **74**), OR
- turn a polyethylene dpm up the wall or back across the top of the flooring.

Ensure that the dpm links with the wall dpc.

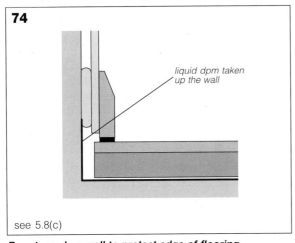

see 5.8(c)

Dpm turned up wall to protect edge of flooring

(d) Where there is no dpm above the floor, protect the edge of flooring from moisture in the wall with a vapour control layer and turn up the wall or back across the floor as described in 5.8(c) (Figures **75** to **77**).

(e) Seal the skirting to the flooring to prevent moist air from within the building reaching the structural floor. Sealing the floor is the most effective way of avoiding the small risk of condensation within the construction (Figures **75** to **77**).

TIMBER BASED FLOORING

5.8 (f) Position the vapour control layer, preferably between the floor panels and the insulation (Figures **76** and **77**). Since composite panels with plastics insulation have a high vapour resistance, it is acceptable to locate the vapour control layer beneath the insulation (Figure **75**). If a liquid vapour control layer is used, ensure that the surface of precast flooring is free from dust and loose material.

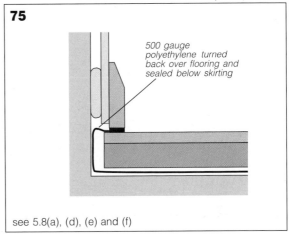

see 5.8(a), (d), (e) and (f)

Vapour control layer beneath composite panels

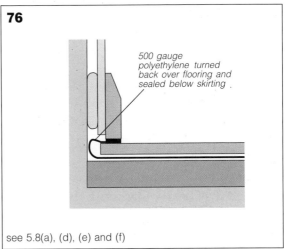

see 5.8(a), (d), (e) and (f)

Vapour control layer within a loose laid system sealed at the skirting

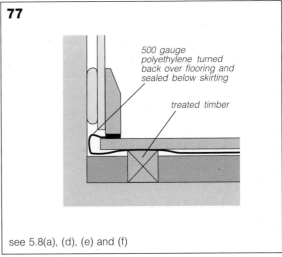

see 5.8(a), (d), (e) and (f)

Vapour control layer within a timber battened system sealed at the skirting

Thermal insulation: avoiding risks

Floors

Concrete ground floors insulated above the structure

5.8 (g) Seal with a flexible sealant or tape around any services that penetrate the damp proof membrane or vapour control layer. Ensure that all holes made when fixing services are sealed.

(h) Limit the potential for damage to timber based flooring panels by using moisture resistant types. For chipboard, use Type C4 to BS 5669; for orientated strand board (OSB), use Type F2; for cement bonded particle board, use Type T2; for plywood, use a type with WBP adhesive.

SCREED FINISH

(i) If there is no dpm above the concrete floor (pre-cast or in-situ), place a vapour control layer of 500 gauge (0.12 mm) polyethylene between the insulation and the screed to protect moisture sensitive finishes (Figure **78**).

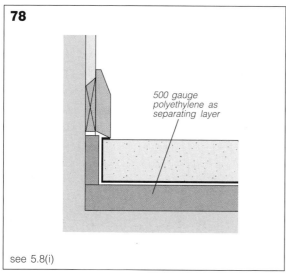

Vapour control layer between screed and insulation

(j) Allow the screed to dry to an appropriate moisture content before the floor finish is laid. Guidance for the requirements for different floor finishes is given in BS 5325, BS 8201 and BS 8203.

5.9 Condensation at thermal bridges

There is a potential thermal bridge at the junction of the floor and the external wall and, where insulation is omitted, around service pipes or ducts.

5.9 (a) Achieve insulation continuity by using this form of floor insulation in conjunction with internal wall insulation, either timber frame or insulated internal lining (Figure **79**).

(b) Start cavity wall insulation at, or below, the bottom of concrete slabs or suspended precast concrete floors, or down to the toe of concrete rafts. Additionally, the use of insulating blocks for the inner leaf of the wall or for precast flooring built into the wall minimises the risk of thermal bridging. The use of high density blocks in these circumstances is not recommended (Figures **80**, **81** and **82**).

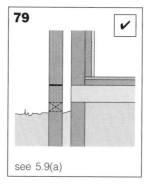
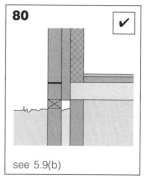
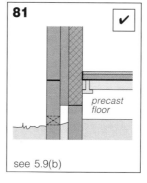
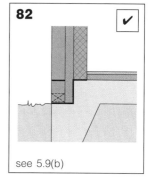

TIMBER BASED FLOORING

(c) Butt insulation boards against the external wall, except when using composite flooring panels which require an expansion gap. This gap should not significantly increase the risk of condensation from thermal bridging.

SCREED FINISH

(d) Insert a vertical strip of insulation at the perimeter of the floor – the full depth of the screed on top of the floor insulation – to avoid the thermal bridge between the screed and the masonry wall. Choose an insulation thickness which ensures that the strip is covered by the wall finish and skirting.

Concrete ground floors insulated above the structure — **Floors**

5.10 Risks associated with services

Water supply pipes below floor insulation can be damaged by frost. Services within the floor can be damaged by fixings. They can also cause dampness when they puncture damp proof membranes or vapour control layers. Some water services need to be accessible for replacement or repair. PVC sheathing to cables can have reduced life expectancy if in direct contact with expanded polystyrene insulants.

5.10(a) Ensure that damp proof membranes and vapour control layers are sealed if penetrated by services or fixings for services (Figure **83**).

(b) Insulate cold water supply pipes throughout their length if they pass through a ventilated subfloor irrespective of their distance from the external wall (Figure **83**). For the recommended insulation thickness, refer to 5.5(b) and Appendix B.

(c) Make a rodent proof, gas tight seal where the water supply pipe enters the building using expanding foam with cement mortar pointing (Figure **83**).

(d) Check all connections and drain valves for tightness before filling pipework with water to avoid damage to insulated flooring.

TIMBER BASED FLOORING

5.10(e) Run water supply pipes in a duct within the insulation thickness and provide a cover for access.

(f) Insulate central heating pipes to concentrate heat output at the heat emitters. Run the pipes in ducts within the insulation thickness, with access only at pipe bends and joints.

(g) Seal around all pipes penetrating the floor to prevent water from leaks or drain valves collecting below the flooring.

(h) Run gas pipes on top of the structural floor within the insulation thickness.

(i) Run electric cables in conduit to avoid contact between PVC sheathing and polystyrene insulation and to limit mechanical damage. Route the conduit within the insulation thickness and specify metal conduit if the cable is within 50 mm of the underside of timber based flooring. Cables enclosed by insulation, especially those serving cooker points and high output heaters, may need to be de-rated (increased in size), even when in conduit (see Appendix A).

SCREED FINISH

(j) Run water supply pipes within the screed in a duct with an accessible cover.

(k) Insulate central heating pipes to concentrate heat output at heat emitters and cover them with at least 25 mm of well compacted screed. Reinforcement or extra cover may be needed in areas with heavy traffic.

(l) Run gas pipes below the screed, within the insulation.

(m) Run electrical cables in conduit to avoid direct contact with polystyrene insulation. Route conduit within the insulation and de-rate the cables as necessary (see Appendix A).

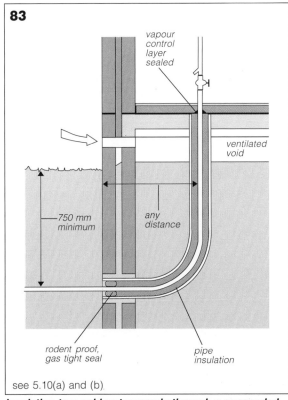

Insulation to a cold water supply through a suspended ventilated floor

Floors

Concrete ground floors insulated at the edge

CHARACTERISTICS OF THE CONSTRUCTION

Insulation is concentrated only at the edge of the floor and positioned either horizontally or vertically for a distance of up to 2 metres, depending on the size and shape of the building and the insulation thickness.

When placed vertically, it can be located:
- on the inside of the external walls,
- within the cavities,
- on the outside of the wall.

Depending on the depth and type of foundation, insulation may be required to the foundation itself, as well as to the wall.

In a few cases, the required U-value may be achieved using only low conductivity solid blockwork for the wall from ground level to the foundation.

The depth of insulation below ground level (D) to achieve a particular U-value can be determined using the calculation method in BRE Information Paper IP 7/93.

R Vertical insulation applied externally may be useful where it is not feasible to upgrade the floor within the building, but it is possible to add insulation to the outside face of the substructure wall if the foundation depth is sufficient. This is particularly useful in conjunction with external wall insulation and it can also be used as a remedy for surface condensation on concrete ground slabs.

| Domestic | ✔ |
| Other sectors | ✔ |

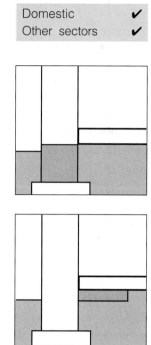

ASSOCIATED TECHNICAL RISKS

- **5.11** Reduced thermal performance of insulating materials below ground
- **5.12** Damage during construction
- **5.13** Structural failure of the substructure
- **5.14** Condensation at thermal bridges

RISKS AND AVOIDING ACTIONS

5.11 Reduced thermal performance of insulating materials below ground

If the insulation absorbs moisture or is degraded by contaminants in the ground or fill, its thermal performance will be reduced.

5.11 (a) Choose an insulant with very low water absorption characteristics. If it is to be in direct contact with the ground or fill, choose a type which will not react chemically with any contaminants present in the ground or fill.

(b) Use a low conductivity block, which has low water absorption, sufficient structural strength and is resistant to frost and ground contaminants, for solid blockwork edge insulation (Figure **84**).

(c) Choose a cavity insulation which is free draining or closed cell and provide drainage to remove water from the cavity at foundation level (Figure **87**).

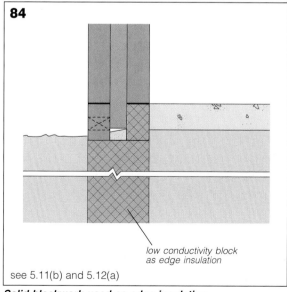

see 5.11(b) and 5.12(a)

Solid blockwork used as edge insulation

Concrete ground floors insulated at the edge — **Floors**

5.12 Damage during construction

Since the insulation is installed in conjunction with groundworks, it may be damaged by machinery or building operations in general.

5.12 (a) Use an insulant which has sufficient strength to resist damage from the process of concreting or trench backfilling.

The greatest resistance to damage is provided by solid blockwork, provided it has sufficient thermal resistance to act as edge insulation (Figure **84**).

(b) Fix internal vertical edge insulation to the face of masonry walls so that it is held firmly in position before backfilling against it.

It may be necessary to insulate around strip footings or trench fill, depending on the required depth of insulation **D**, calculated by the method in BRE IP 7/93 (Figure **85**).

(c) Protect external vertical edge insulation below dpc level by a board, render on metal lath, or paving slabs.

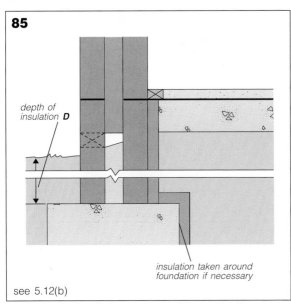

see 5.12(b)

Vertical edge insulation on internal face of wall

5.13 Structural failure of the substructure

The introduction of thermal insulation materials into the substructure can affect the structural stability of the ground slab and of the external walls below ground level.

5.13 (a) Ensure that the support given to the concrete slab by the horizontal strip of insulation and the remainder of the floor slab is constant.

(b) If horizontal edge insulation is set into the sand blinding, the same slab thickness can be used across the whole floor (Figure **86**).

(c) If horizontal insulation is laid above the dpm or level sand blinding, increase the overall slab thickness to ensure that the minimum required thickness is maintained.

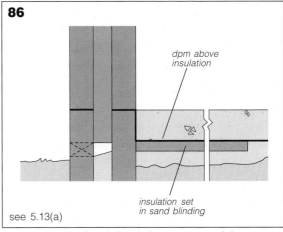

see 5.13(a)

Horizontal edge insulation below concrete slab

(d) When vertical edge insulation is placed in a cavity wall down to foundation level, ensure that inner and outer leaves are tied together at 450 mm above foundation level and thereafter to suit the coursing of the internal blockwork and the cavity wall insulation. Use stainless steel ties below ground level (Figure **87**).

(e) Thicken the inner support leaf of a cavity wall which is to be insulated to foundation level, if the distance from the slab to the foundation is more than 600 mm and it cannot be guaranteed that the trench will be backfilled and compacted evenly in stages on each side of the wall (Figure **87**).

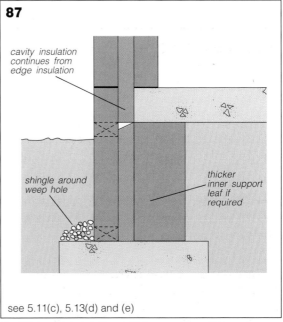

see 5.11(c), 5.13(d) and (e)

Vertical edge insulation within a cavity wall

Floors
Concrete ground floors insulated at the edge

5.14 Condensation at thermal bridges

Where there are breaks in the continuity of the insulation layers, a thermal bridge is created and surface condensation can occur.

5.14 (a) Avoid a thermal bridge at the junction of wall and floor by:
- achieving continuity of insulation in the same plane, ie internal vertical edge insulation with internal wall insulation, cavity edge insulation with cavity wall insulation and external edge insulation with external wall insulation (Figures **89** to **91**), OR
- overlapping insulation layers and linking them with insulating blockwork (Figures **92** and **93**).

(b) Ensure that the edge of the screed is insulated when it links internal wall insulation and internal vertical edge insulation Figure **91**).

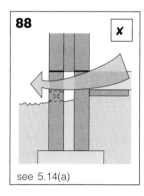
88 ✗
see 5.14(a)

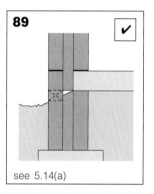
89 ✓
see 5.14(a)

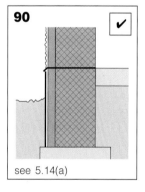
90 ✓
see 5.14(a)

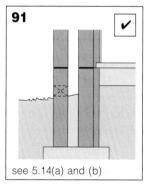
91 ✓
see 5.14(a) and (b)

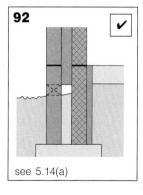
92 ✓
see 5.14(a)

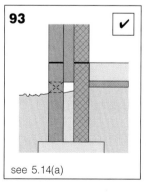
93 ✓
see 5.14(a)

Thermal insulation: avoiding risks

Suspended timber ground floors

Domestic ✔
Other sectors ✔

CHARACTERISTICS OF THE CONSTRUCTION

The advantage of this type of floor is that the required thickness of insulation can be accommodated easily within the thickness of the structure, even allowing for thermal bridging at the joists.

The insulation materials may be mineral wool quilt supported on netting or boards or rigid plastics boards supported on timber battens.

R Relevant when the flooring and possibly the joists of an existing suspended timber floor need to be renewed or where insulation can be fitted below the floor.

ASSOCIATED TECHNICAL RISKS

5.15 Increased heat loss due to air movement and reduced insulation thickness
5.16 Condensation at thermal bridges
5.17 Damage from moisture
5.18 Risks associated with services

RISKS AND AVOIDING ACTIONS

5.15 Increased heat loss due to air movement and reduced insulation thickness

The heat loss through the floor is increased if cold air is able to enter the building from the subfloor space, if gaps in the insulation layer allow air movement above the insulation or if insulation is reduced in thickness.

5.15 (a) Use full depth strutting between the joists at the perimeter of the floor to provide a fixing for the flooring and to limit the gaps which need to be sealed to those at the edge of the floor (Figure **94**).

(b) Seal the skirting to the wall with a sealant, and to the floor with a flexible sealant or an extruded draughtproofing section. Also seal gaps at services penetrations and access panels in the flooring (Figure **94**).

(c) Pack the space between the wall and the joist or solid strutting with mineral wool insulation (Figure **94**).

(d) Ensure that the space between plasterboard dry lining and masonry walling is sealed by applying continuous ribbons of plaster adhesive at the perimeter of the wall, particularly at skirting level (Figure **94**).

(e) Position the insulation level with the top of the joists to avoid any air movement between the insulation and the flooring (Figures **94** and **96**), unless the movement of cold air from the subfloor

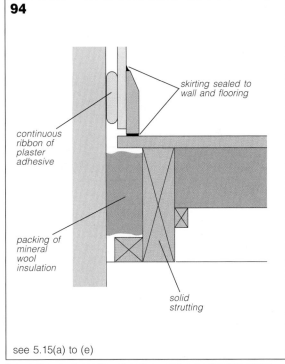

see 5.15(a) to (e)

Insulating and sealing the floor at the perimeter of the floor

space through or around the insulation is restricted, eg by using support boards for quilt installation (Figure **95**) or tightly butted plastics insulation boards (Figure **97**).

Floors
Suspended timber ground floors

QUILT INSULATION

5.15(f) Avoid draping insulation over the joists as its thickness will be reduced substantially either side, and the uneven support will lead to loose fixings and a squeaky floor.

(g) Choose a support method which ensures that the full thickness of insulation is maintained for the full width between the joists.

(h) Support quilt insulation, either:
- on a board, fixed to battens nailed to the sides of the joists, to reduce air movement through and above the insulation (Figure **95**), OR
- on plastics netting or strips of vapour permeable breather membrane when using large tongued and grooved flooring panels which themselves create an airtight layer. Fine netting may be draped over the joists and held against the sides with staples or battens (Figure **96**).

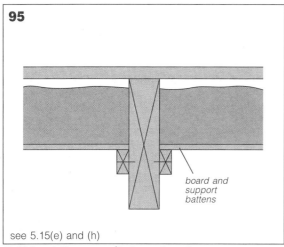

95

board and support battens

see 5.15(e) and (h)

Quilt insulation supported on a board

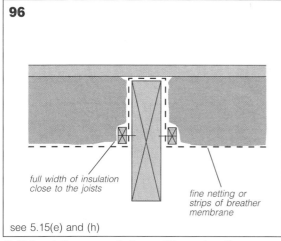

96

full width of insulation close to the joists

fine netting or strips of breather membrane

see 5.15(e) and (h)

Quilt insulation supported on netting or breather membrane

PLASTICS BOARD INSULATION

5.15(i) Support rigid plastics boards on battens fixed to the sides of the joists and hold the boards down against the battens with nails driven into the joists (Figure **97**).

Do not support insulation boards on nails or clips, unless the insulation is tight to the underside of the flooring, since any gap due to the inaccurate cutting of the boards will allow cold air to by-pass the insulation.

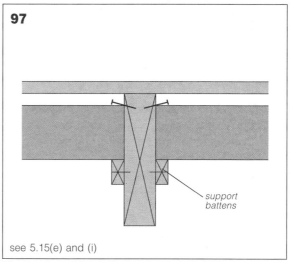

97

support battens

see 5.15(e) and (i)

Plastics board insulation supported on battens

5.16 Condensation at thermal bridges

Condensation can occur on the surface of the wall or floor where there is a thermal bridge between the inside of the building and the cold subfloor space. This can be via the external wall or the edge of the floor.

5.16(a) Maintain insulation continuity by packing the space between the external wall and the joist or solid strutting with mineral wool insulation which can be cut conveniently to suit the space available. It will be an advantage to set out the first joist, or position solid strutting, at least 50 mm from the wall (Figures **98** and **99**).

(b) Use insulating blockwork in the thermal bridge path between the inside of the building and the subfloor space (Figure **99**), or choose a wall specification with an insulated internal lining.

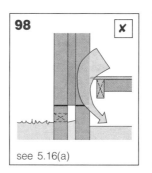

98 ✗

see 5.16(a)

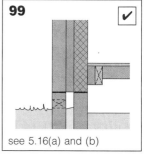

99 ✓

see 5.16(a) and (b)

Suspended timber ground floors **Floors**

5.17 Damage from moisture

Adequate cross ventilation of the subfloor space is necessary to keep timber and timber based products dry, as they can be damaged by moisture. Insulation below the flooring can restrict the space available for cross ventilation.

5.17 (a) Cross ventilate the subfloor space with ventilators in at least two opposite walls, with a minimum free area of 1500 mm² per metre run of wall, or a total free area of 500 mm² per square metre of ground floor area, whichever is the greater.

(b) Ensure that ventilators are not obstructed by floor insulation.

(c) Where strutting is needed above sleeper walls, use herringbone strutting since solid strutting limits the ventilation area to that through the honey-combed sleeper wall.

(d) Position floor insulation so that any ventilation provision over sleeper walls will not be obstructed.

(e) It is not necessary to introduce a vapour control layer into insulated timber suspended floors, since any small amounts of condensation that form will be safely ventilated away.

(f) Limit damage to timber based flooring panels by using moisture resistant types. For chipboard, use Type C4 to BS 5669; for orientated strand board (OSB), use Type F2; for cement bonded particle board, use Type T2; for plywood, use a type with WBP adhesive.

5.18 Risks associated with services

Services within the floor can be damaged by fixings, frost, overheating or chemical reaction. Water supply pipes below floor insulation can be damaged by frost. Services within the floor can be damaged by fixings. Some water services need to be accessible for replacement or repair. PVC sheathing to cables can have reduced life expectancy if in direct contact with expanded polystyrene insulants. Holes for services reduce the airtightness of the floor.

5.18 (a) Insulate cold water supply pipes as they pass through the ventilated subfloor, as described in 5.10(b).

(b) Run central heating pipes above the floor insulation and insulate the pipes to concentrate heat output at the heat emitters.

(c) Ensure that the floor insulation does not block ventilation paths to solid fuel appliances which are designed to receive their air supply from the subfloor space.

(d) Provide access at 2 metre intervals to cold water supply pipes which run horizontally within the floor using access panels in large timber based flooring panels, or use removable floor boards.

(e) Run gas pipes below the insulation, between or below the joists.

(f) Avoid contact between PVC sheathed cables and polystyrene insulation, or run the cables in conduit.

(g) Use cables with an earthed metallic sheath or run the cables in metal conduit when they are closer than 50 mm to the top of the joist. Cables enclosed by, or close to floor insulation, especially those serving cooker points and high output heaters, may need to be de-rated (increased in size), even when in conduit (see Appendix A).

(h) Seal around services where they pass through the flooring.

Floors

Concrete and timber upper floors

Domestic ✔
Other sectors ✔

CHARACTERISTICS OF THE CONSTRUCTION

Upper floors can be intermediate floors between flats or maisonettes, or floors whose soffit is exposed to the outside air. This section does not specifically apply to floors above unheated or ventilated spaces, although the principles outlined still apply.

Concrete floors are constructed in-situ, or use precast units. Floor slabs often have edge beams supported from masonry walls. Junctions with the external wall and where slabs project at balconies are the most difficult to detail.

Intermediate floors may need to be insulated to meet sound insulation requirements. Exposed floors always need to be insulated, either above, within or below the structure, to meet thermal requirements.

The insulation methods and products applicable to upper floors are often the same as those used for concrete and timber ground floors, but techniques similar to those used for external insulation are also available when insulating below concrete soffits, or for flat roofing when insulating the tops of balconies.

R The insulation of intermediate and exposed floors is common when existing buildings are being converted or renovated. Buildings with a concrete frame with masonry infill present particular problems.

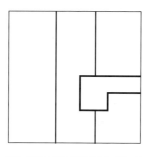
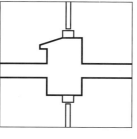
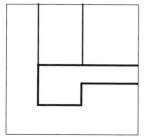

ASSOCIATED TECHNICAL RISKS
5.19 Condensation at thermal bridges
5.20 Condensation within the construction

RISKS AND AVOIDING ACTIONS

5.19 Condensation at thermal bridges

When a concrete intermediate floor has an edge beam built into a cavity external wall, the edge beam often projects into the cavity and the continuity of cavity insulation is broken, causing a thermal bridge. If the floor projects to the outside face of the wall, or beyond it to form a balcony, the insulation layer is also interrupted. Similarly, edge beams supporting the external wall at exposed floors create a potential thermal bridge and the risk of condensation.

INTERMEDIATE FLOORS
5.19(a) Project wide concrete beams into the room, not into the wall cavity, to ensure continuity of cavity insulation (Figures **100** and **101**). A projection may be acceptable where the beam is covered with at least 50 mm of cavity insulation and the insulation is protected by a cavity tray (Figure **102**).

R (b) Use cavity insulation in conjunction with insulation on the face of the edge beam where circumstances permit, eg where the beam partially projects into the outer leaf and was previously finished with brick slips (Figure **103**).

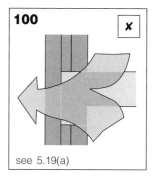
see 5.19(a)

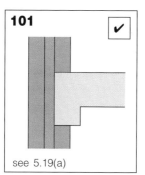
see 5.19(a)

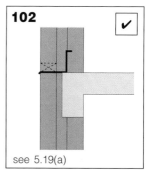
see 5.19(a)

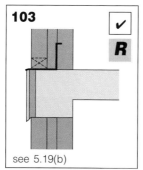
see 5.19(b)

Concerete and timber upper floors **Floors**

5.19(d) Where cavity insulation is interrupted or cavities extend over several storeys, take account of published guidance on:
- cavity trays and weep holes,
- damp proofing,
- fixings to concrete beams,
- allowance for movement in masonry,
- cavity barriers,
- support for the outer masonry leaf (Figure **104**).

(e) Avoid continuity from the cold supporting flange of the metal angle, through its vertical flange, to the concrete beam by fixing the angle to vertical support channels, with insulation behind the metal angle and between the support channels (Figure **104**).

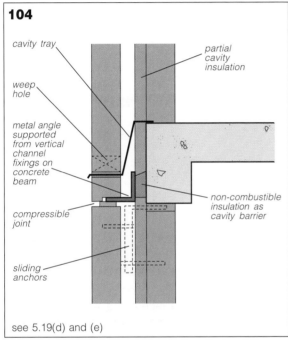

see 5.19(d) and (e)

Detailing at edge beams with partial cavity insulation

(f) Use external insulation in preference to internal insulation, since the insulation layer is not broken by the floor slab or edge beam.

R (g) Insulate with external insulation or insulated overcladding when concrete beams (and columns) are flush with the face of the external wall (Figure **105**).

(h) If internal insulation is used in conjunction with cavity masonry walls, insulate the edge beam within the cavity and on internal surfaces (Figure **106**).

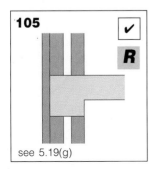

see 5.19(g)

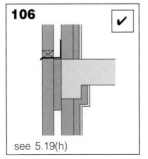

see 5.19(h)

BALCONIES

5.19(i) Where possible, design balconies to be structurally separated from the floor slab, so that the insulation can be continuous through the floor (Figure **108**).

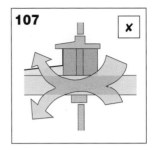

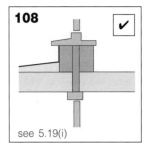

see 5.19(i)

(j) Where the floor and balcony slab is continuous, use cavity or external insulation and insulate the soffit of the slab within the heated space with a strip of insulation with a thermal resistance of at least 0.6 m^{2}K/W for a distance of at least 300 mm from the window/door frame or wall (Figures **109** and **110**).

It is an advantage if the insulation is cast into the slab so that it extends above the door or window frame.

R (k) Where the floor and balcony slab is continuous, follow the guidance in 5.19(j), but surface fix the insulation and its protective finish (Figures **109** and **110**).

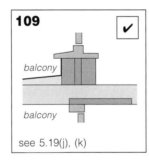

see 5.19(j), (k)

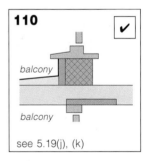

see 5.19(j), (k)

(l) When rooms for occupation are above a balcony or access deck:
- insulate the entire soffit of the balcony, OR
- insulate the floor of the occupied room and the ceiling of the room below as in 5.19(j).

Use a vandal-resistant finish to insulation applied externally (Figures **111** and **112**).

R (m) When rooms for occupation are above a balcony or access deck, follow the guidance in 5.19(l), using surface fixed insulation internally (Figures **111** and **112**).

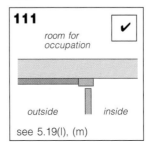

see 5.19(l), (m)

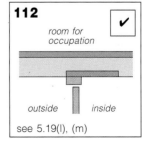

see 5.19(l), (m)

Thermal insulation: avoiding risks

Floors — Concrete and timber upper floors

EXPOSED FLOORS

5.19(n) For exposed floors, ensure that:
- there is continuity of external wall and floor insulation, OR
- insulating blockwork links overlapping layers of insulation (Figures **113** to **116**).

(o) Support the external leaf of cavity walls independently and insulate below concrete slabs and around edge beams to ensure insulation continuity (Figure **114**).

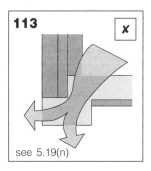
113 ✗
see 5.19(n)

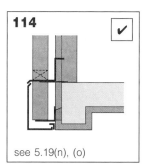
114 ✓
see 5.19(n), (o)

(p) When external wall insulation is used:
- extend it around the edge beam to link with insulation below the slab, OR
- extend it down the face of the edge beam and use insulating blockwork to link with insulation above the slab (Figures **115** and **116**).

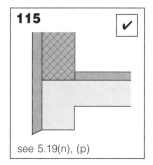
115 ✓
see 5.19(n), (p)

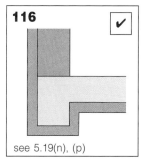
116 ✓
see 5.19(n), (p)

(q) Where the external leaf is supported by a concrete edge beam which is a continuation of the floor slab, use insulating blockwork in combination with cavity insulation or internal insulation and insulate above the concrete slab (Figures **117** and **118**).

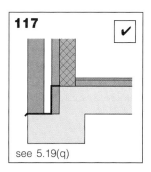
117 ✓
see 5.19(q)

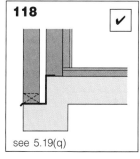
118 ✓
see 5.19(q)

5.19(r) For timber floors, ensure that the space between the wall and the joist or strutting is fully insulated (Figures **119** to **122**).

(s) To avoid a thermal bridge at the junction between insulated timber floors and edge beams:
- use cavity insulation in conjunction with an inner leaf of insulating blockwork, OR
- extend external wall insulation around the edge beam to achieve continuity with the floor insulation, OR
- use internal wall insulation (Figures **120** to **122**).

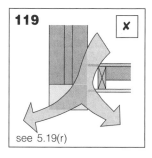
119 ✗
see 5.19(r)

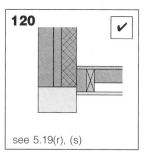
120 ✓
see 5.19(r), (s)

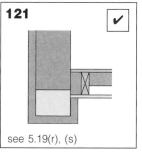
121 ✓
see 5.19(r), (s)

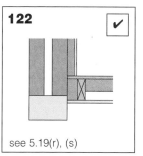
122 ✓
see 5.19(r), (s)

5.20 Condensation within the construction

When internal linings, including insulated linings, cover cold surfaces, such as concrete edge beams and floor slabs close to the exterior, moisture from within the building can condense on the concrete and damage the insulation and/or lining.

Where exposed floors are insulated below the structure, condensation can occur within the floor if water vapour in the internal air cannot permeate through the construction.

5.20(a) At intermediate floors and balconies, ensure that internal linings to edge beams and adjacent ceilings are sealed or incorporate a vapour control layer to prevent moist air reaching cold concrete surfaces.

(b) For exposed soffit floors:
- ensure that the vapour resistance of materials above the insulation is at least five times greater than that of materials below the insulation, (eg the soffit lining), OR
- if this is not possible, provide a 50 mm ventilation space behind the soffit lining, with openings at opposite edges of the soffit, equivalent to a continuous 25 mm gap.

Quality control checks for floors

CONCRETE GROUND FLOORS
- Is the dpm correctly positioned and continuous with the wall dpc?
- Is insulation tightly fitted and, when in contact with the ground, is it rigid, of low water absorption and, where necessary, resistant to chemicals?
- Are the insulation and dpm undamaged by concreting or screeding?
- Is insulation at the edge of the floor installed when necessary to reduce thermal bridging?
- Is the cold water supply pipe insulated and the dpm sealed where penetrated by the pipe?
- Are there minimal services run beneath the flooring and are those that are, accessible for maintenance if required?
- Are liquid dpm materials chemically compatible with expanded polystyrene insulation?
- Is there a gap of at least 10 mm between timber based flooring panels and the wall or rigid upstands?
- Is screeding over insulation carried out in accordance with the specification?

SUSPENDED TIMBER GROUND FLOORS
- Is insulation supported directly beneath the flooring without gaps and is it packed into spaces at the floor edge between the wall and the joist or solid strutting?
- Are services correctly located and insulated as specified and is the floor insulation reinstated after installation of services?
- Is underfloor ventilation clear and not restricted through and over the sleeper walls and at perimeter wall ventilators?
- Is plasterboard bonded to masonry walling with a continuous ribbon of plaster adhesive at skirting level and is a flexible sealant applied between the flooring and the skirting and between the skirting and the plasterboard?

CONCRETE AND TIMBER UPPER FLOORS
- Are special provisions followed as specified for reducing thermal bridging at the perimeter of intermediate floors, balconies and exposed floor slabs?

Thermal insulation: avoiding risks

Appendix A

Sizes for cables enclosed within thermal insulation

The presence of insulation around a cable has the effect of reducing the current carrying capacity and, in some cases, this will require the cable to be increased in size to safely carry the load.

In general, the effect of cables being enclosed by insulation is as follows:

- circuits run within thermal insulation must be protected with cartridge fuses or mini-circuit breakers (MCBs). Rewirable fuses are not suitable.
- cables fully enclosed by insulation may need to be increased in size above the standard recommended size by as much as 20% if they pass at right angles through an insulating layer, and as much as 50% if they are enclosed along their length for more than 500 mm.
- for cables enclosed by insulation but in contact with a thermally conductive surface on one side, the larger of the standard recommended sizes will generally need to be used.

Specific guidance on the effect on cables of thermal insulation in pitched roof spaces is given in BS 5803: Part 5.

Each case should be separately calculated by reference to the IEE Wiring Regulations 16th Edition, BS 7671 or the Electricians' Handbook (1993 edition) available from the Electricity Association.

The Electricians' Handbook contains correction factors and tables of current carrying capacities according to whether a cable is fully enclosed by insulation (installation method 1) or covered by insulation but in contact with a thermally conductive surface, eg plasterboard, on one side (installation method 4).

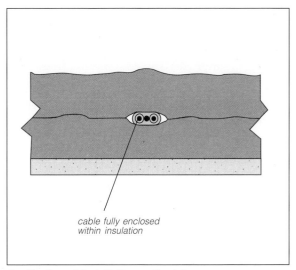

Application of installation method 1

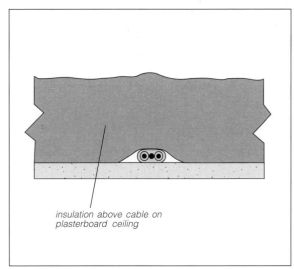

Application of installation method 4

Appendices

Appendix B

Insulation thickness for water supply pipes (based on the revision to BS 6700)

Nominal outside diameter of pipe [mm]	Minimum thickness of insulation to delay freezing [mm][1]			
	Thermal conductivity (λ) of insulating material at 0°C[2]			
	0.035 W/mK	0.040 W/mK	0.045 W/mK	0.055 W/mK
15	25	32	45	70
22 to 28	19	22	27	32
35 and over	9	13	19	25
Flat surface	9	13	19	25

Notes

(1) The thickness of insulation under the respective thermal conductivity values is considered reasonable to provide a minimum of 8 hours' protection in normally occupied buildings. Absence in excess of 24 hours is not considered normal occupation. In unoccupied buildings, and where severe weather conditions prevail, additional protection will be required.

(2) Thermal conductivity when tested in accordance with BS 874. For polyethylene circular pipe insulation, guidance on conductivity tests can be obtained from the Polyethylene Foam Insulation Association.

(3) The conditions assumed for the table are:
- ambient temperature $-6°C$
- water temperature $+7°C$
- ice formation   50%.

Appendix C

Sources of information

British Standards Institution

BS 476 *Fire tests on building materials and structures.*

BS 5250 *Code of basic data for the design of buildings: the control of condensation in dwellings.*

BS 5618 *Code of practice for thermal insulation of cavity walls (with masonry or concrete inner and outer leaves) by filling with ureaformaldehyde (UF) foam systems.*

BS 5628 *Code of practice for use of masonry.* Part 3 *Materials and components, design and workmanship.*

BS 6232 *Thermal insulation of cavity walls by filling with blown man-made mineral fibre* Part 1 *Specifications for the performance of installation systems.*

BS 6676 *Thermal insulation of cavity walls using mad-made mineral fibre batts (slabs)* Part 2 *Code of practice for installation of batts (slabs) filling the cavity.*

BS 6700 *Specification for design, installation, testing and maintenance of services supplying water for domestic use within buildings and their curtilages.*

BS 8104 *Code of practice for assessing exposure of walls to wind driven rain.*

BS 8208 *Guide to assessment of suitability of external cavity walls for filling with thermal insulants* Part 1 *Existing traditional cavity constructions.*

Building Research Establishment

BRE Defect Action Sheet 109 *Hot and cold water systems – protection against frost.*

BRE Digest 369 *Interstitial condensation and fabric degradation.*

BRE Digest 377 *Selecting windows by performance.*

BRE Digest 380 *Damp-proof courses.*

BBA, BEC, BRE, NHBC

Cavity insulation of masonry walls – dampness risks and how to minimise them.

BDA, BCA, BPF, ACBA, EURISOL, NCIA

CIW–2 *Cavity insulated walls – specifiers' guide.*

BFRC and CIRIA

Flat roofing: design and good practice.

British Steel

Colourcoat in building: a guide to architectural practice.

Department of the Environment (EEO), BRECSU

Good Practice Guides 93 to 97 *Detailing for designers and building professionals* and 98 to 110 *Site practice for tradesmen.*

Good Practice Guide 61 *Design manual: energy efficiency in advance factory units.*

Electricity Association

Electricians' Handbook.

Glass and Glazing Federation

Glazing techniques for insulating glass units.

Institution of Electrical Engineers

Regulations for electrical installations.

London Brick Company

Climate and brickwork: constructional notes.

National House-Building Council

NHBC Standards.

Zurich Municipal

Zurich guarantees technical manual.

Index

Air admittance valves, 4, 17
Air movement
    dry lining to walls, 26, 27, 55
    partial cavity insulation, 26
    sarking insulation, 8
    suspended timber ground floors, 55
    window/wall junctions, 37

Balconies, 59
Bathrooms, 4, 5, 27
Breather membranes, 5, 8, 31

Cavity barriers
    fire, 6, 18, 26, 31, 59
Cavity closers, 6, 18, 35, 36
Cavity trays, 23, 24, 59
Cavity wall insulation, 1, 23, 24, 54, 58, 60
Chipboard flooring, 47–50, 57
Cladding
    insulated overcladding, 31
    masonry walls, 23, 24, 27
Cold bridging (see Thermal bridging)
Composite panels
    flooring, 47–50
    insulating plasterboard, 27, 28, 32
    sheet cladding, 31
    sheet roofing, 10, 12
Compression of insulation
    concrete ground slab, 44
    screeds, 47, 48
    timber based flooring, 47, 56
Condensation
    cisterns, 7
    cold deck flat roofs, 17
    double glazing units, 39
    double windows, 41
    extract ducting, 5, 9
    floors, exposed soffit, 60
    floors, timber ground, 57
    internal surfaces, 1
    internal wall insulation, 27
    lightweight metal decking, 15
    metal window frames, 38
    pipes, 7, 9
    pitched roofs, 4, 5
    profiled sheet roofing, 10
    rainwater downpipes, internal, 12, 14, 16, 18
    secondary glazing, 41
    summer condensation, 28
    (see also Interstitial condensation and Thermal bridging)
Conduit
    electrical, 29, 33, 46, 51, 57
Construction moisture,
    concrete ground floors, 46
Corrosion
    profiled sheet roofing, 12
Cracking
    plaster, 26
    render, 30

Damp proof membranes
    damage to, 44
    position in floor, 43, 44, 46, 49
    sealing when punctured, 46, 51

Delamination
    profiled sheet insulated panels, 12, 31
Double glazing units
    edge seals, 39
    glazing methods, 40
Double windows, 41
Draught seals
    loft hatch, 4
    skirtings, 49, 55
    window/wall junctions, 37
Driving rain, 21
Dry lining to walls
    air movement behind, 26, 27, 55

Electrical cables
    contact with expanded polystyrene, 7, 29, 33, 51, 57
    overheating, 7, 29, 33, 51, 57, 63
    sizes of cables, 63
Electrical fittings
    sealing gaps around, 4
    recessed in ceiling, 7
Expanded polystyrene
    contact with electrical cables, 7, 29, 33, 51, 57
    warm deck flat roof, 13
Exposure of walls, 21
External wall insulation, 1, 30, 54, 59, 60
Extract ducts
    insulation, 5, 9
Extract ventilation, 1, 2, 10, 27

Fire hazards
    combustible insulation, 13, 28, 30
    inability to escape through windows, 41
    metal deck roofs, 13
    overheating cables, 7, 29, 33, 51, 57, 63
    wall cavities, 26, 28, 31
Flat roofs
    cold deck, 17
    inverted warm deck, 15
    sandwich warm deck, 13
Floors
    balconies, 59
    concrete, 43–54, 58–60
    edge insulation, 52
    quality control checks, 61
    timber, 55–57, 58–60
Floor finishes
    failure in service, 47
    moisture sensitive, 46, 49, 50
Flue outlets
    location, 4, 17
Freezing
    pipes and cisterns, 7, 46, 51, 57, 64
Frost
    attack on walls, 25
    resistance of blockwork below dpc, 52

Glazing
    double glazing units, 39
    secondary glazing, 41

Thermal insulation: avoiding risks

Index

Heating
    appropriate systems, 1
    controls, 1, 2
    insulating distributing pipework, 51, 57

Insulating blockwork
    below dpc level, 45, 52
    cavity closer, 6, 18, 35
    shrinkage, 26

Insulation
    central heating pipes, 51, 57
    cisterns, 7
    damage during construction, 13, 48, 53
    degradation, 16, 44, 52
    extract ducts, 5
    internal rainwater downpipes, 12, 14, 16, 18
    resistance to ground contaminants, 44, 52
    sarking, 8
    support in timber ground floors, 56
    warning pipes, 7
    water absorption, 44, 52
    water supply pipes, 46, 51, 57, 64

Internal wall insulation, 1, 27–29, 36, 54, 59, 60

Interstitial condensation
    impervious wall cladding, 26, 31
    internal wall insulation, 27
    inverted warm deck flat roofs, 16
    profiled sheet roofing, 10
    sandwich warm deck flat roofs, 14
    sarking insulation, 8
    timber based flooring, 49
    timber framed walls, 32
    upper floors, 60

Joints
    window/wall junctions, 37

Kitchens, 5, 27, 48

Lighting
    low energy, 2
Lintels, 36
Loft hatch, 4, 7
Loft insulation, 3

Masonry walls
    cavity, 25
    protection to, 23, 24
    rain penetration, 21
    solid, 23, 30
Membranes
    breather, 5, 8, 31
    damp proof, 43, 44, 46, 49, 51
    (see also Vapour control layers)

Overcladding, 31

Painted masonry, 25, 28
Pipes
    freezing, 7, 46, 51, 57, 64
    position in floor, 46
    sealing around, 4, 46, 51
Pitched roofs
    profiled sheet, 10
    sarking insulation, 8
    room-in-the-roof, 5
    ventilated lofts, 3
Plaster cracking, 26

Plasterboard dry lining
    air movement behind, 26, 27, 55
Profiled sheet
    cladding, 31
    roofing, 10

Quality control checks
    floors, 61
    roofs, 19
    walls, 34
    windows, 42

Rain penetration, 21
Render, 23, 24, 30
Roofs
    cold deck flat, 17
    flat, 13–18
    inverted, 15
    pitched, 3–12
    profiled sheet, 10
    quality control checks, 19
    room-in-the-roof, 5
    warm deck flats, 13

Screeds
    ducts for services, 51
    failure in service, 47
    insulation below, 48, 50, 54
    moisture content, 50
Separating walls, 29, 45
Services
    ground floors, 46, 51
    roof spaces, 7
    sealing around, 4, 27, 55, 57
    timber framed walls, 32, 33
Sheet cladding
    delamination, 12, 31
    fire risk, 31
Shrinkage
    blockwork, 26
Stack vents
    termination, 4, 17
    thermal movement, 4, 17
Stop-ends, 23, 24
Sulphate attack
    masonry cavity walls, 25
Summer condensation, 28

Thermal bridging
    balconies, 59
    exposed floors, 60
    flat roofs, 14, 16, 18
    ground floor/wall junctions, 45, 50, 54, 56
    lintels, 36
    pitched roofs, 6, 9
    profiled sheet roofing, 11
    separating wall junctions, 29, 45
    walls, 25
    window/wall junctions, 35–37
    upper floor/wall junctions, 58–60
Thermal capacity, 1, 2
Thermal movement
    stack vents, 4, 17
Timber based flooring
    moisture resistance, 49, 50, 57
    provision for expansion, 47
    support, 47, 48, 56

Index

Timber floors
  suspended ground floors, 55–57
  upper floors, 58–60
Timber framed walls, 1, 32, 33

UV degradation
  of insulation, 16

Vapour barriers (see Vapour control layers)
Vapour checks (see Vapour control layers)
Vapour control layers
  cold deck flat roofs, 17
  internal wall insulation, 27
  profiled sheet roofing, 10
  room-in-the-roof, 5
  sandwich warm deck flat roofs, 14
  screeds, 50
  sealing when punctured, 5, 27, 32, 51
  summer condensation, 28
  timber based flooring, 49
  timber framed walls, 32
  upper floors and balconies, 60
Ventilation
  cold deck flat roofs, 17
  double windows, 41
  ducting in roof spaces, 5
  exposed soffit linings, 60
  extract ventilation, 1, 2, 10, 27
  impermeable cladding, 31
  mechanical, 2, 27
  pitched roofs, 4, 5
  secondary glazing, 41
  suspended timber floors, 57

Walls
  cavity, 25
  external insulation, 1, 30, 54, 59, 60
  internal insulation, 1, 27–29, 36, 54, 59, 60
  masonry, 21, 23–25, 30
  quality control checks, 34
  separating, 29, 45
  timber framed, 1, 32, 33
Warning pipes
  insulation, 7
Weatherproof membrane
  abrasion by grit, 16
  fatigue, 13
Wind
  scouring gravel ballast, 16
Windows
  condensation channels, 38
  deformation of frames, 40
  double glazed, 39
  draught seals, 37
  metal framed, 38
  quality control checks, 42
  thermal breaks, 38